W0225806

HIGH PRESSURE METHODS
IN
SOLID STATE RESEARCH

Springer Science+Business Media, LLC

©
Springer Science+Business Media New York
1969
Originally published by Butterworth & Co. (Publishers) Ltd in 1969
Softcover reprint of the hardcover 1st edition 1969

ISBN 978-1-4899-5879-2   ISBN 978-1-4899-5877-8 (eBook)
DOI 10.1007/978-1-4899-5877-8

Suggested U.D.C. number: 621–186.5
Suggested additional number: 66.083.2
Library of Congress Catalog Card Number: 68–58922

ACKNOWLEDGMENTS

I would like to express my gratitude to Mr. N. B. Owen of the National Physical Laboratory for reading and commenting on a large part of the manuscript and particularly for his help in preparing Appendices A and B.

I would like to thank also my wife, Vivien, for her assistance and encouragement.

C. C. Bradley

CONTENTS

PAGE

Acknowledgments v

1. Introduction 1
2. Construction Materials for High-Pressure Apparatus .. 15
3. Hydrostatic Pressure Apparatus 24
4. Opposed Anvil Apparatus 52
5. Multi-anvil Devices 87
6. Piston and Cylinder Apparatus for Pressures up to 100 Kilobars 122
7. Miscellaneous Methods 142

Appendix A. Materials Commonly Used in High-Pressure Apparatus 168

Appendix B. Suppliers of Materials and Equipment in the United Kingdom 171

Index 173

1

INTRODUCTION

In recent years the use of the pressure parameter in materials research has increased enormously. It is probably true to say that the growth rate in the 1960s compares favourably with that of liquid helium temperature research in the 1950s. It has been estimated that there are of the order of three hundred laboratories actively engaged in making high pressure measurements, and nearly six hundred papers on the subject were published in 1965. By far the greatest growth has been in the United States and the National Bureau of Standards has set up a data centre at Brigham Young University, Utah, to compile all published high pressure work. In other countries the growth rate has been at a lower level, although in the USSR the Institute for High Pressure Research of the Academy of Sciences, Moscow, has been to the forefront in this field for many years. Probably the most important event has been the synthesis of diamond at high temperature and high pressure by General Electric Co., (U.S.A.) and by A.S.E.A. Co. (Sweden) and its wide economic consequences. The basis of the most important techniques in high pressure research were worked out by the late P. W. Bridgman and the debt to him will be obvious from the descriptions given in the remainder of this book.

Before discussing briefly the significance of the degree of pressure it is useful to define the units which are commonly used. These are the bar (b) and kilobar (kb), the atmosphere (atm), kilograms per square centimetre (kg/cm^2) and pounds per square inch (lb/in^2). The unit adopted in this book is the *bar* (or *kilobar*) in common with the vast majority of high pressure researchers. The equivalence of the units is given below.

1 Bar $\equiv 10^6$ dynes/cm^2 $\equiv 0.9869$ atmospheres (normal)
$\equiv 1.0197$ kg/cm^2
$\equiv 14.504$ lb/in^2

It can be said that almost any experiment capable of being carried out at ordinary room conditions can be performed at 10 kb with modern advances in sophistication of high-pressure techniques. Fermi surface contours have been measured in single crystals up to 8 kb at 4.2°K and nuclear magnetic resonance experiments in

1

the 60 kb region are possible. This is not the upper limit by any means, but as the pressure range is widened effects arising from the method of transmitting pressure to a material become increasingly important. Above approximately 30 kb at room temperature hydrostatic pressure media cannot be used since they freeze and they are replaced by soft solids which have small but significant shear moduli and hence may introduce inhomogeneous forces. Above 80 or 90 kb the absolute pressure scale is not well established, although individual measurements referred to an arbitrary scale can be made quite accurately.

The present limit of static pressure experiments is about 500 kb and the only available method for producing pressures above this is by shock wave techniques. Since this method is beyond the scope of the normal solid state research laboratory it has not been considered in this book. There are several comprehensive reviews and the reader is referred to them for further information[1,2].

A number of books have been published during the last four or five years containing review articles on a wide variety of phenomena at high pressure and it would be pointless (and virtually impossible) to attempt a summary. Some of them are listed at the end of the chapter.

The purpose of this book is to give the interested reader an insight into the design and limitations of high-pressure apparatus over the whole range of static pressures to 500 kb. Bridgman's classic book *The Physics of High Pressure* (published by G. Bell and Sons, 1958) contains several chapters on techniques dealing almost exclusively with the hydrostatic range. Wentorf's *Modern Very High Pressure Techniques* contains more modern methods concentrating exclusively on the range above 30 kb. The design and construction of the apparatus described in later chapters is well within the capabilities of the technical services of most research laboratories but in any case there are many small engineering firms anxious to provide a service in this field.

It is not the aim of the book to provide a complete review of all the methods which have been devised and for the most part original versions are chosen for description since modifications of these are to a large extent the individual efforts of a number of different experimenters.

Many of the devices described in detail in subsequent chapters have been built at the National Physical Laboratory and used by the author and his colleagues. Suitable construction materials are given in Chapter 2 and Appendix A and are basically those which can be obtained easily in most scientifically developed countries.

Emphasis has been placed on the use of well-established materials, for example nickel steels, since although there may be available new materials of preferred properties these may not have been tested over a large enough period to justify unqualified recommendation. Specifications, particularly for steels, have been given, and the particular compositions are given in Appendix A.

The question of safety precautions in high-pressure experiments should not be understated. There is considerable stored energy in materials at high pressure particularly in gases and liquids and it is important at all stages to incorporate adequate safety screens around apparatus in which high pressure is being generated.

PRESSURE TRANSMISSION AND MEASUREMENT

As has been stated previously this book is concerned exclusively with the problem of generating and measuring static pressures. The range 0–500 kb is conveniently divided at around 30 kb into two regions for the purpose of discussing pressure transmission and calibration. In the low pressure range 0–30 kb the pressure media used usually are fluids and a sample is subjected to truly hydrostatic pressures with no shear effects. In the higher range solid transmitting media are employed resulting in varying degrees of uniformity of pressure. In this case there is always a finite shear stress, albeit quite small in solids like silver chloride and boron nitride, but it can effect the onset of phase transitions in some cases. In the range 0–30 kb pressure is usually monitored continuously with either a Bourdon tube (below 5 kb) or manganin resistance manometers, these being calibrated against given fixed points or free piston gauges. In the higher range the normal procedure is to determine the applied load versus pressure relation at a few fixed points and to extrapolate. The applied load then serves as a measure of the pressure. In the case of piston and cylinder apparatus fairly sensible friction corrections can be made to load over area values but in the case of compressible gasket apparatus this is not possible and measurements depend heavily on the extrapolations. This latter method has to be used since continuously recording manometers such as the resistence gauge would be subjected to large uncertainties arising from inhomogeneous straining. In spite of this, they have been used by a few researchers for this pressure region.

Many laboratories have investigated pressure measurement in detail and it forms a major undertaking in high pressure research.

However, the aim here is to give fairly convenient methods which are reliable within the limits stated.

0–30 kb Range

The most fundamental method of measuring pressure in fluid systems is to balance against a column of mercury. This is limited to pressures of a few hundred bars with a few exceptions and is inconvienent to use.

A second primary standard up to 26 kb is the dead weight or free piston gauge. In this the fluid pressure is balanced by a piston loaded with weights. There are several versions available and accuracies up to 0·1 per cent can be obtained. These are really only convenient in a laboratory which is well set up to make this kind of measurement (Chapter 3).

The most convenient method is to use a manganin resistance manometer (Chapter 3). This was first used extensively by P. W. Bridgman who showed that the resistance versus pressure relation is linear to 0·1 per cent up to 12 kb and to 2 per cent at 25 kb[3]. Manganin resistance gauges are very easy to make and to use

Table 1.1

Material	Nature of discontinuity		Pressure, kb	Reference
Mercury	Freezing point at	0°C	7·569 ± 0·001	18
	Freezing point at	+20°C	11·54 ± 0·02	19, 20
	Freezing point at	−20°C	2·677 ± 0·002	20
Carbon tetrachloride	Freezing point at	20°C	3·310 ± 0·010	20
Bismuth I–II	Solid–Solid	25°C	25·38 ± 0·08	21
Bismuth II–III	Solid–Solid	25°C	26·97 ± 0·20	21
Bismuth III–V*	Solid–Solid	25°C	89 ± 2	24
			78 → 82	25
Caesium I–II	Solid–Solid	25°C	22·6 ± 0·6	21
Caesium II–III	Solid–Solid	25°C	41·7 ± 1·0	21
Thallium II–III	Solid–Solid	25°C	36·69 ± 0·2	21
Barium I–II†	Solid–Solid	25°C	58·5 ± 0·5	7
			59·2 ± 1·0	22
Tin*	Solid–Solid	25°C	114	23
Iron*	Solid–Solid	25°C	133	23
Barium II–III*	Solid–Solid	25°C	144	23
Lead*	Solid–Solid	25°C	160	23
Rubidium*	Solid–liquid(?)	25°C	193	23
Calcium*	Solid–Solid	25°C	375	23

* These points to be treated with caution (see text).
† Recently Haygarth et al.[26] have redetermined the Barium I–II point in a modified single stage piston and cylinder apparatus. Their value of 55·0 ± 0·5 kb (at 25°C) is below that normally quoted. Hence this fixed point should be regarded now as tentative until there is further evidence from other sources.

4

although differences up to 1 per cent in the resistance/pressure slope may occur in winding different coils from the same spool. If properly heat and pressure cycled initially they are very reproducible with negligible hysteresis effects. Two calibration fixed points are usually used, the freezing pressure of mercury at 0°C (7·569 kb) and the I–II transition in bismuth at 25·4 kb (*Table 1.1*). Alternatively calibrations against a free piston gauge can be made up to 26 kb. The accuracy of the resistance gauges is obviously increased by enlarging the number of fixed points and it has been suggested that other points found in *Table 1.1* are used[4]. However for most purposes an accuracy of about 0·3 per cent is quite adequate and is obtained easily with two calibration points. The effects of temperature are neglible up to 35°C but beyond this point care should be exercised in pressure measurements. Further details of manganin manometers are given in Chapter 3. Other resistance gauges such as gold–chrome have proved to be not as reproducible as manganin and hence are not used widely.

30–90 kb Range

The method of calibration in this range, where pressure is usually transmitted by soft solids, is by a number of 'fixed' points resulting from polymorphic transitions in a number of common substances. Generally a load versus pressure relation is obtained by extrapolation between the fixed points and subsequent measurements are made by reading off at the appropriate load value. (Recently a free piston gauge usable to 100 kb has been built by Vereshchagin et al.[7] but it has not been used extensively as yet.)

The use of polymorphic transitions as calibration points arose from the work of Bridgman[5,6] on pressure/volume and pressure/electrical resistance relations in a large number of elements and compounds. A number of phase changes in the range up to 100 kb proved to be both easily measured and reproducible. These together with sharpness are the main requirements for a fixed point. As yet it is not possible in general to compare on an absolute basis with free piston gauges and the best compromise is to measure the transitions in a piston and cylinder apparatus and to make corrections to the load over area pressure value for the friction and bore distortion. This was the procedure adopted by Bridgman and more recently by Kennedy and La Mori[8]. Unfortunately the limitation of piston and cylinder apparatus means that calibrations above 60 kb are not easily made and any fixed point in this range has to be regarded as tentative.

INTRODUCTION

Originally two pressure scales were established, one based on Bridgman's volume transitions in a piston and cylinder, and the other on his electrical resistance measurements in opposed anvils. Except for the lowest bismuth point at 25 kb the two scales deviate by more than 50 per cent at the higher end, assuming that the particular phase changes are the same in each case. It has been shown, however, that pressure generated in opposed anvils is not simply force over area as assumed by Bridgman (see Chapter 4), and the two scales were brought into coincidence by simultaneous measurements of volume and resistance. In other words the scale based on the volume discontinuities is the most reliable one and is in fact the one recommended by Bridgman. Kennedy and La Mori redetermined some of the transitions and their values are usually quoted for an absolute scale between 25 and 60 kb (*Table 1.1*).

In practice, resistance measurements are more easily made since they require only small volumes of the relevant substance and can be monitored with greater accuracy. This is especially important in pressure apparatus in the 30–500 kb region where volumes are on the small side especially those over which quasi-hydrostatic pressures are maintained. In many cases it is possible to have a calibrant material present in an experiment together with the sample under investigation, which would be much less convenient generally with a volume discontinuity measurement.

A list is given below of the most commonly used transitions although it by no means exhausts all possibilities. They have in common a fairly large discontinuity in volume and/or resistance, good reproducibility and occur in materials which are comparatively easy to obtain and handle. They are summarized in *Table 1.1*.

Bismuth I–II, II–III and III–V transitions—The nomenclature used for these transitions follows Bridgman's work on the bismuth phase diagram and the early confusion between the resistance and volume scales lead to a different nomenclature for the III–V transition. In addition Bridgman found a volume but no resistance transition at 45 kb which has not been confirmed by other people. Normally the III–V transition at 89 kb is known as the 'upper' bismuth point. Its value, based on Bridgman's volume work, may be inaccurate and should be treated accordingly[24].

Thallium II–III transition—Kennedy and La Mori[8] have shown that the Bridgman figure of 40 kb was incorrect and the accepted value is now 36·7 kb.

Caesium II–III transition—This falls in a useful range between the thallium II–III and barium I–II transition, but caesium is difficult to handle and is not used extensively.

Barium I–II transition—The usually accepted value for this based on Bridgman's work is 59 kb but recent redeterminations have given values of 53 ± 1kb[9] and 55·0 ± 0·5 kb[26]. It is difficult to measure absolutely since it lies just at the limit of the straightforward single stage piston and cylinder (Chapter 6). Barium metal oxidizes rapidly in air and special techniques have to be employed to ensure good electrical contact.

Typical resistance vs. pressure curves are shown in *Figure 1.1*.

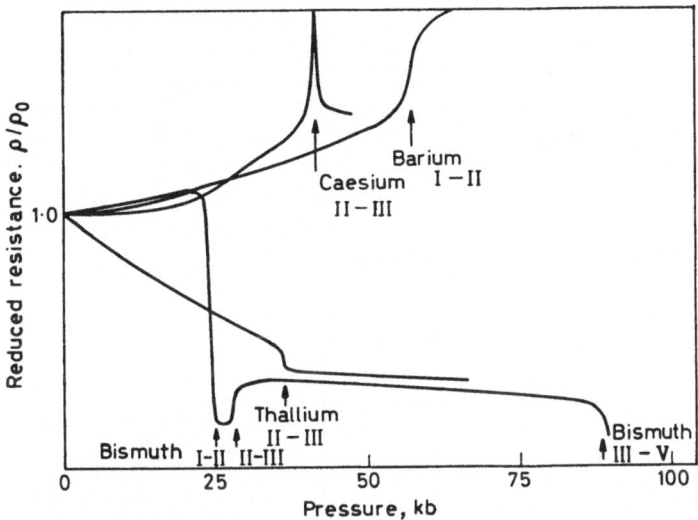

Figure 1.1. Resistance vs. pressure for calibration metals Bi, Th, Cs and Ba

These fixed points are always subject to review although the bismuth I–II and II–III should be reasonably reliable since they can be measured in hydrostatic media. For this reason all reported experimental work should quote the fixed points used in the pressure scale. Work published before 1960 and based on the resistance scale of Bridgman is still useful as long as the appropriate corrections to the scale are made. This may be necessary in present work if the fixed points above are redetermined at different values by later workers[9,26].

90–500 kb Range

Above 90 kilobars there are a number of fixed points (listed in *Table 1.1*) which are based on resistance discontinuities in lead, iron, barium, rubidium, calcium and tin. These have been measured by

Drickamer and others[10] in supported anvil apparatus and rely on extrapolation of an electrical resistance scale in platinum beyond 89 kb and are for this reason somewhat uncertain. The iron 131 kb point agrees well with shock wave work but not so well with x-ray measurements by Jeffery *et al.*[9] who redetermined the upper bismuth point on which these points are based on a new scale (see below).

These transitions are obviously useful but the actual pressures at which they occur should be regarded as tentative.

Pressure Calibration at Elevated Temperatures

Calibrations carried out at room temperature are often not applicable at high temperatures since there is frequently a redistribution of stress especially in solid cell assemblies. The normal procedure is to follow an established phase line in a *P–T* diagram for a convenient substance. The diamond/graphite and quartz/coesite boundaries have been used but they are more difficult than those based on simple melting curves.

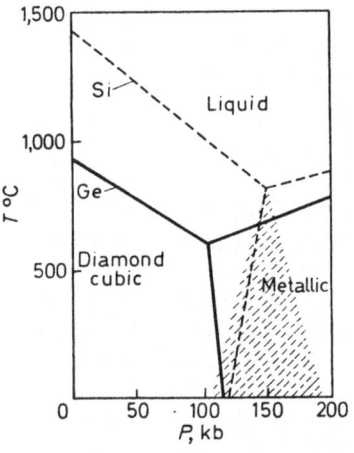

Figure 1.2. Phase diagram for Si and Ge, corrected for pressure effect on thermocouple readings and local pressure due to local heating (after Bundy[13])

The latter have been determined for a number of pure elements and they afford a good basis for measuring pressures at elevated temperatures. The melting curves of germanium and silicon measured by Bundy[13] are shown in *Figure 1.2;* others may be obtained from the work of Kennedy and Newton[11,12]. Melting points are usually monitored quite easily by noting arrests on thermocouple readings on heating or cooling.

Pressure Scale Based on the Equation of State for Sodium Chloride

The equation of state has been determined for sodium chloride over the range 0–500 kb and 0–1500°C by Jeffery *et al.*[9], from the work of Decker[14]. It is suggested that by measuring the volume of sodium chloride under given conditions an absolute pressure scale can be established. Although there are some reports of phase changes in sodium chloride around 20 kb which would eliminate its possibility in this use these have not been confirmed by other people[9]. The volume of sodium chloride has been measured by x-ray diffraction at high temperature and high pressure and it has been shown that the barium I–II and bismuth III–V transitions may be 10 per cent less than the presently accepted values on this scale. Further confirmation of the equation of state is necessary before it can be accepted universally.

Temperature Measurement at High Pressures

The most commonly used method for monitoring temperature in high pressure apparatus is by a thermocouple. Resistance thermometers are generally too bulky and suffer from inaccuracy due to inhomogeneous straining.

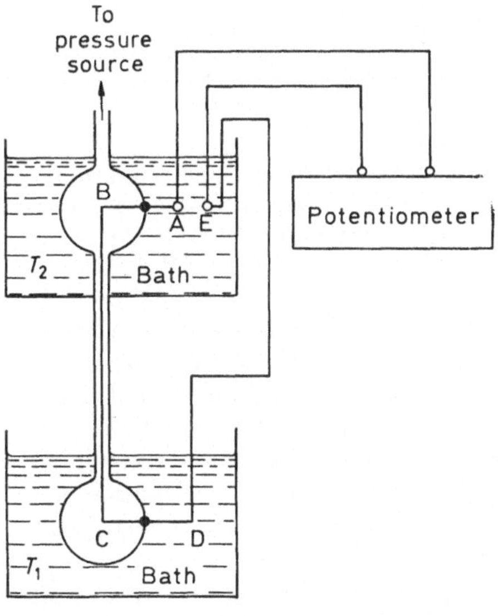

Figure 1.3. Bridgman's apparatus for measuring pressure e.m.f. of metals. A, B, C, D, E is metal under test; AG and EF are copper (after Bridgman[15])

The electromotive force developed in a thermocouple depends on the electronic properties of the metals concerned. These are changed

9

by the application of pressure and therefore thermocouple readings should be corrected from atmospheric pressure calibrations when used in high pressure apparatus. There has been some research in this field. Bridgman[15] measured the effect of pressure on the thermoelectric power of a metal in the arrangement shown in *Figure 1.3*. Essentially the same metal is used but that part under pressure is considered different for thermoelectric purposes.

The pressure range was extended to 70 kb by Bundy[16], using the arrangement shown in *Figure 1.4* in a high-pressure belt apparatus (Chapter 6). In the configuration in *Figure 1.4a* there is no difference

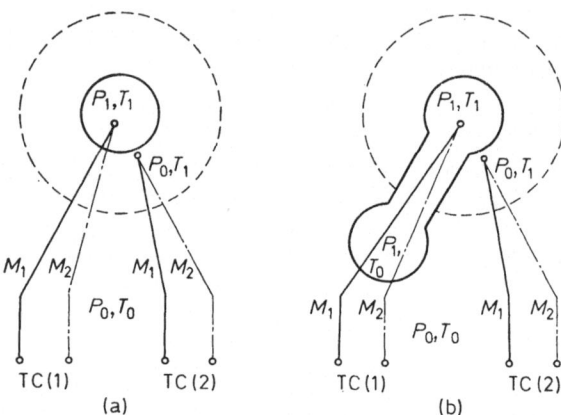

Figure 1.4. (a) Totally heated pressure apparatus. (b) Experimental arrangements required for measuring pressure effect on thermocouples by direct comparison (after Bundy[16])

between the readings of the two thermocouples. (They are assumed to be identical in construction, that is, the second represents in effect the atmospheric calibration of the first.) This is because where there is a pressure change the conditions are isothermal. In *Figure 1.4b* there is a difference since now the temperature gradient is wholly in the high-pressure region. The measurements of Bundy are of limited use since the temperature range was only 100°C. A few examples are shown in *Figure 1.5*.

Hanneman and Strong[17] have extended the temperature range to 1,300°C by using a different technique. From work on diamond synthesis it has been shown that the Simon–Berman graphite/diamond equilibrium line obtained by calculation from thermodynamic data is quite accurate. Using a belt apparatus and calibrating pressures at elevated temperatures from the germanium melting curve it is possible to obtain a pressure/temperature relation for diamond synthesis where the temperature is that apparently

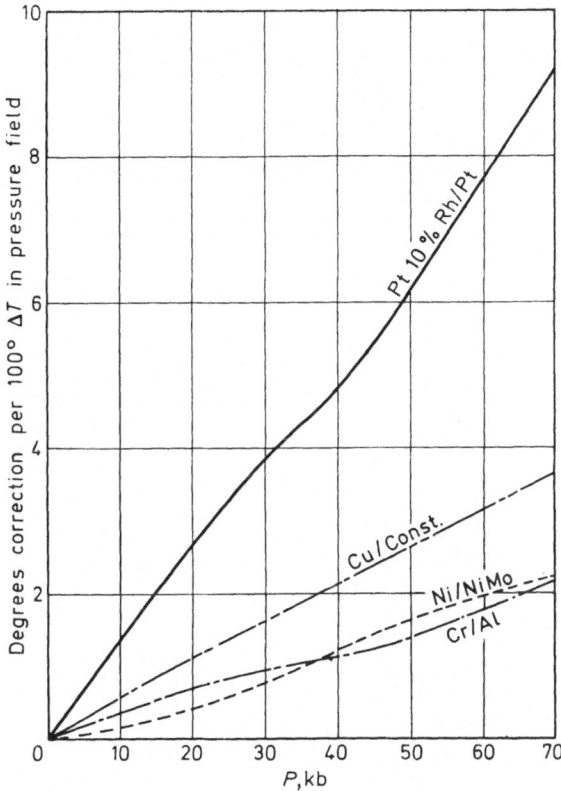

Figure 1.5. Pressure corrections to four common thermocouples for ΔT of 100°C (after Bundy[16])

recorded by a given thermocouple. Comparison with the Simon–Berman line furnishes a correction to the thermocouple reading. The results from this method agree closely with those obtained in a similar manner from diffusion experiments and the α–γ phase diagram in iron. *Figure 1.6* shows the relative corrections for three common thermocouples. It will be seen that by far the smallest

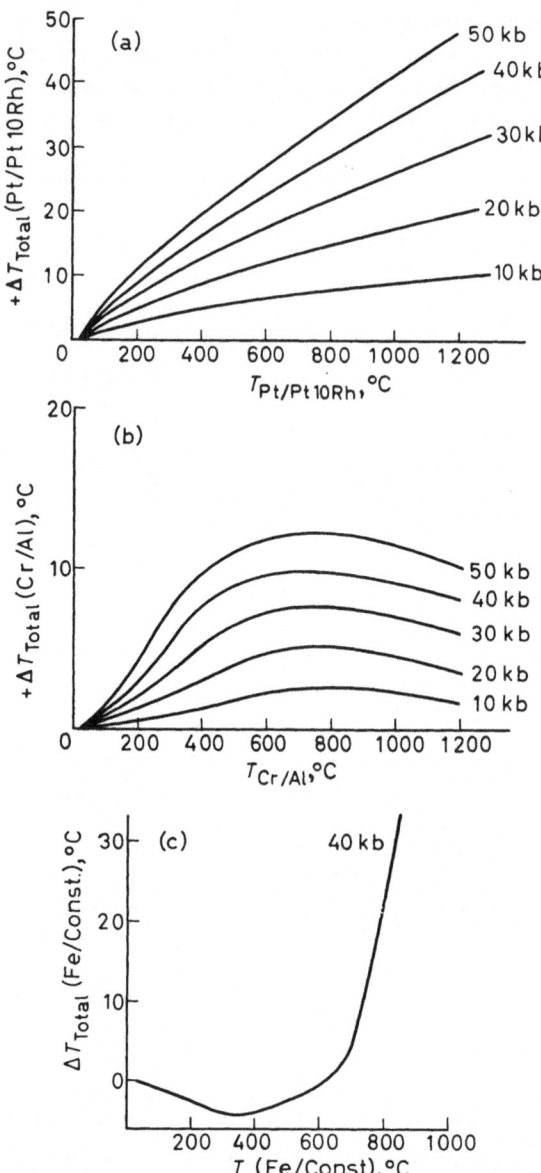

Figure 1.6. $\Delta T_{\text{Absolute}}$ vs. observed T at various isobars for (a) Pt/Pt 10 Rh, (b) chrome/alumel, and (c) iron/Constantan thermocouples (after Hanneman and Strong[17])

effect is obtained with chromel–alumel. The accuracy is of the order of $\pm$ 20 per cent in the ΔT values.

REFERENCES

[1] Wentorf, R. H. (Ed.). *Modern Very High Pressure Techniques.* Butterworths, London, 1962.

[2] Hamann, S. D. *Advances in High Pressure Research,* vol. 1, p. 85. Ed. R. S. Bradley. Academic Press, New York, 1966.

[3] Bridgman, P. W. *Proc. Amer. Acad. Arts Sci.* 1912, **47**, 321.

[4] Boren, M. D. Babb, S. E., jr., and Scott, G. J. *Rev. sci. Instrum.* 1965, **36**, (10), 1456.

[5] Bridgman, P. W. *Proc. Amer. Acad. Arts Sci.* 1942, **74**, 425.

[6] Bridgman, P. W. *Proc. Amer. Acad. Arts Sci.* 1952, **81**, 167.

[7] Vereshchagin, L. F., Zubova, E. V., Buimova, I. P. and Burdina, K. P. *Sov. Phys. Dokl.* 1966, **11**, 585.

[8] Kennedy, G. C. and La Mori, P. N. *Progress in Very High Pressure Research,* p. 304. Eds. F. P. Bundy, Hibbard and H. M. Strong. Wiley, New York, 1961.

[9] Jeffery, R. N., Barnett, J. D., Vanfleet, H. B. and Hall, H. T. *J. appl. Phys.* 1966, **37**, 3172.

[10] Balchan, A. S. and Drickamer, H. G. *Rev. sci. Instrum.* 1961, **32**, 308.

[11] Kennedy, G. C. and Newton, R. C. *Solids Under Pressure.* Eds. W. Paul and D. Warschauer, p. 163. McGraw-Hill, 1963.

[12] Jayaraman, A., Klement, W., Newton, R. C. and Kennedy, G. C. *J. Phys. Chem. Solids* 1963, **24**, 7.

[13] Bundy, F. P. *J. chem. Phys.* 1964, **41**, 3809.

[14] Decker, D. L. *J. appl. Phys.* 1965, **36**, 157.

[15] Bridgman, P. W. *Proc. Amer. Acad. Arts Sci.* 1918, **53**, 269.

[16] Bundy, F. P. *J. appl. Phys.* 1961, **32**, 483.

[17] Hanneman, R. E. and Strong, H. M. *G. E. Res. Rep. No. 64-RL-3725X,* 1964.

[18] Dadson, R. S. and Greig, R. G. P. *Brit. J. appl. Phys.* 1965, **16**, 1711.

[19] Babb, S. E. jr. *High Pressure Measurement,* p. 115. Eds. A. A. Giardini and E. C. Lloyd. Butterworths, London, 1963.

[20] Babb, S. E. jr. *Technique of Inorganic Chemistry,* vol. 6, p. 83. Wiley, New York, 1966.

[21] Kennedy, G. C. and La Mori, P. N. *J. Geophys. Res.* 1962, **67**, 851.

[22] Jayaraman, A., Klement, W. and Kennedy, G. C. *Phys. Rev. Letters* 1963, **10**, 387.

[23] Drickamer, H. G. *High Pressure Measurement,* p. 34. Eds. A. A. Giardini and E. C. Lloyd. Butterworths, London, 1963.

[24] Bridgman, P. W. *Proc. Amer. Acad. Arts Sci.* 1948, **76**, 55.

[25] Klement, W., Jayaraman, A., and Kennedy G. C. *Phys. Rev.* 1963, **131**, 632.

[26] Haygarth, J. C., Getting, I. C. and Kennedy, G. C. *J. appl. Phys.* 1967, **38**, 4557.

INTRODUCTION

For Further Reading

Progress in Very High Pressure Research. Edited by F. P. Bundy, W. R. Hibbard, jr. and H. M. Strong. Wiley & Sons, New York, 1960.

Modern Very High Pressure Techniques. Edited by R. H. Wentorf. Butterworths, London, 1962.

High Pressure Physics and Chemistry, vols. I and II. Edited by R. S. Bradley. Academic Press, New York, 1963.

High Pressure Measurement. Edited by A. A. Giardini and E. C. Lloyd. Butterworths, London, 1963.

Solids Under Pressure. Edited by W. Paul and D. Warschauer. McGraw-Hill, New York, 1963.

Physics of High Pressure and the Condensed Phase. Edited by J. Van Itterbeek. North Holland, Amsterdam, 1965.

Physics of Solids at High Pressures. Edited by C. T. Tomiznka and R. M. Emrick. Academic Press, New York, 1965.

Advances in High Pressure Research, vol. I. Edited by R. S. Bradley. Academic Press, New York, 1966.

2

CONSTRUCTION MATERIALS FOR HIGH-PRESSURE APPARATUS

The choice of materials for use in high-pressure apparatus depends on size, operating temperature, cost and availability as well as the more obvious factors of strength and machinability. The most commonly used over the whole pressure range up to several hundred kilobars are a number of alloy steels and cemented tungsten carbide. The mechanical properties of steel vary enormously with composition and heat treatment and hence make it the most versatile material. In comparison tungsten carbide has only limited use but is essential where the highest pressures are to be generated. The remainder of this chapter will be concerned mostly with the properties of steels.

The relevant mechanical properties are usually considered under the following headings. Further details may be obtained from the references cited at the end of this chapter[1,2].

Hardness and Strength

A typical stress versus strain curve for a tensile test on a sample of low-alloy carbon steel is shown in *Figure 2.1*. OA is the elastic

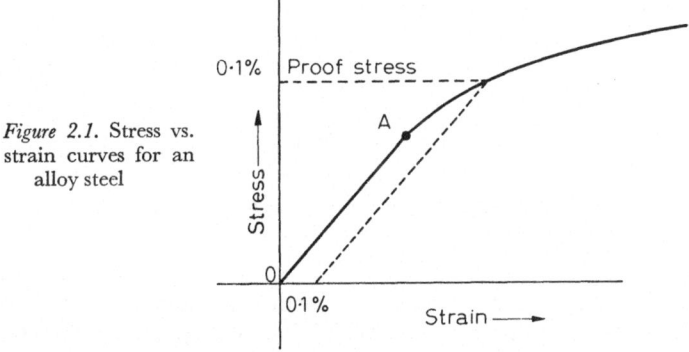

Figure 2.1. Stress vs. strain curves for an alloy steel

region where Hooke's law is operative. Above A the curve rises further due to work hardening and after reaching a maximum it

15

falls again to the breaking point. (Stress is always referred to the load over the original unstressed area and strain as per unit original length.) Operation beyond the limit of the elastic region is usually determined by the proof stress criterion. This is defined as the stress at which there is a given permanent elongation of the bar under test. The normal values are 0·1, 0·2 or 0·5 per cent and are always quoted in data on a particular sample. As shown in *Figure 2.1,* if a line is extrapolated from the proof stress point back to the strain axis parallel to the elastic portion of the stress-strain curve, it cuts this axis 0·1, 0·2 or 0·5 per cent to the right of the origin. The ultimate tensile strength is the highest point reached on the curve. The limit of the elastic region can be raised considerably by work hardening although this has only a small effect on the tensile strength. The relevance of these criteria for high-pressure apparatus will become more clear in the sections on cylinders in Chapters 3 and 6.

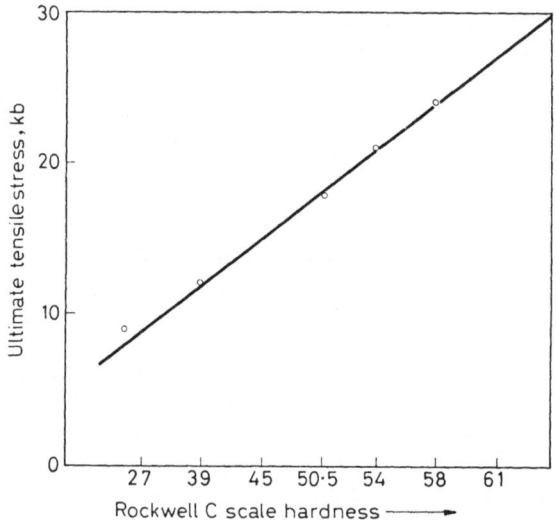

Figure 2.2. Approximate relation between tensile stress and hardness factors (note the non-linearity of the hardness scale)

It is obviously inconvenient to measure directly the proof stress or tensile strength of a material which has been prepared for use in the construction of a device, and instead a hardness test is employed. The term hardness is difficult to define but it has a direct relationship with the tensile strength. The hardness value is conveniently measured by indentation tests with hand-operated machines such as the Brinell, Vickers or Rockwell types. The Rockwell C scale based on tests with a diamond cone indentor with 150 kg·loading is used throughout this book. *Figure 2.2* illustrates an approximate general relation between hardness factors and ultimate tensile strengths for alloy steels.

Elastic Modulus

This is also known as Young's modulus and is determined by the slope of the stress–strain curve over the portion where elastic conditions prevail. Its value is a measure of the degree of distortion under stress and has obvious relevance in the design of a high-pressure container.

Ductility

The ductility of a material is a measure of its resistance to failure under sudden shock. There are several formal definitions of ductility which depend on elongation or diminution of area of a specimen undergoing a tensile test but for the purpose of this account the relevance to brittleness is more important. A steel can be heat-treated to give it a very high tensile strength but at the same time leaving it in a state very susceptible to brittle fracture and most unsuitable for most high-pressure uses. There are a number of tests involving the breaking of notched bar specimens under sudden impact of a load which give an indication on an arbitrary scale of the degree of ductility. The ones commonly quoted are the Charpy and the Izod which differ only in the method of supporting the notched bar and are expressed in terms of footpounds load. It is sufficient for most purposes to say that high impact values indicate good ductility and low ones vice versa; that is, the optimum condition for a steel, for example, would be high tensile strength and high impact test value.

Fatigue

It is usually found that in materials undergoing rapid load cycling failure occurs at stresses only a fraction of the tensile strength due to the phenomenon known as fatigue. This is often difficult to determine with any certainty but in the majority of high-pressure experiments considered in this book it is not relevant enough to cause a great deal of concern, although its consequences are a hazard since it causes sudden brittle failure.

17

Creep

This occurs in materials under stress at elevated temperatures for long periods. Most serious flow or creep takes place above 370°C and the creep rate should be known before using particular steels. Those with small quantities (1 or 2 per cent) of molybdenum or tungsten have most resistance to creep.

The basic requirements of a high-pressure container are strength or hardness and ductility and the optimum combination is found in a number of alloy steels. A part of the iron–carbon phase diagram for low carbon content steels is shown in *Figure 2.3*. The γ and α-phases correspond respectively to face-centred and body-centred

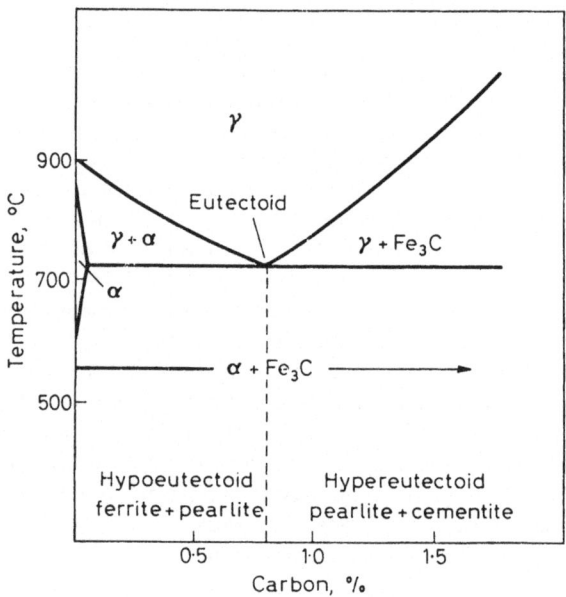

Figure 2.3. An important part of the iron–carbon phase diagram (after Comings[3])

cubic structures of iron containing carbon in solution. The γ-phase is known as austenite and the α-phase as ferrite. If a particular composition is cooled down from the γ-phase to temperatures still above that of the eutectoid the composition consists of austenite together with either cementite (Fe_3C), if the percentage of carbon is greater than 0·8, or α-iron solution with carbon if it is less than 0·8. On further cooling the austenite transforms to ferrite and carbon

18

is rejected from solution in the form of cementite. The mixture of these two is known as pearlite. Steel cooled by this means has a very low strength and is referred to usually as annealed. However, if the cooling is carried out rapidly so that the austenite does not transform at the indicated temperature a room-temperature composition of unstable supersaturated carbon solution in a distorted α-iron lattice called martensite is formed which is extremely hard and brittle. If this is subsequently heated to a temperature below the γ-α transition point and held there for some time before cooling some of the brittleness is annealed out and the result is a material of high strength and good ductility. This procedure is known as tempering.

The important process is the rapid quenching which in most ordinary carbon steels cannot be rapid enough, except in small pieces, to prevent the formation of pearlite. Alloying the steels with small percentages of other metals improves the hardening process so that a more extensive and uniform strength is obtained. The actual hardness depends mostly on the percentage of carbon with the alloying elements having only second-order effects. Water quenching is often more effective in terms of degree of hardening but since it sometimes leads to cracking, oil quenching or air cooling is employed instead.

Another important aspect is the grain size in the steel which is created during the manufacturing process. The best results are obtained with steel which has been remelted in vacuum so that included gases are removed. This results in a smaller grain structure and correspondingly higher strength and less probability of flaws.

A few of the more common alloy steels are given below.

Nickel Alloy Carbon Steels

The alloy steels commonly employed in the construction of apparatus described in this book are nickel chromium molybdenum steels and in particular the British Standard designated EN 24, 25, 26 and 30 B types. The actual compositions are listed in *Table 1* in Appendix A together with a brief summary of their properties. The ultimate tensile strength and Izod impact values (on an arbitary scale) for EN 26 are shown in *Figure 2.4* for a range of tempering temperatures after oil quenching from 850°C. The machinability of a piece of steel depends on hardness and composition but as a general rule the rough machining should be done before heat treatment, and finished by grinding after tempering. The hardness of alloy steels is increased by cold working with subsequent stress-relieving anneals to obtain stability. When this is carried out with high-pressure cylinders it is known as autofrettage (page 26).

EN 30 B contains more nickel (4·25 per cent) than EN 25 or 26 which increases its hardenability. It is therefore used where large sizes are needed but in other respects it behaves in a similar manner to the other nickel steels. The ultimate tensile strength of these steels is of the order of 16 kb with a reasonable ductility. This corresponds to a Rockwell C scale value of about 50.

These nickel alloy steels cannot be used at very low temperatures since they undergo a ductile to brittle transition. The most common use is for low-pressure cylinders and for supporting rings.

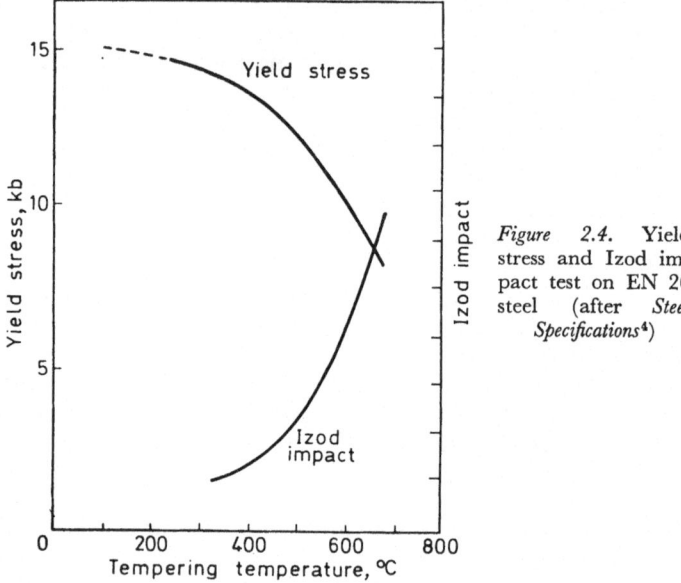

Figure 2.4. Yield stress and Izod impact test on EN 26 steel (after *Steel Specifications*[4])

Tool Steels

These are alloy steels with higher carbon content than those described above. They are hardened to higher strengths but tend to be more brittle. The compositions of three examples given in *Table 1*, Appendix A are taken from the Edgar Allen range. Further details are given in Appendix A.

K 9—This is hardened at 780–800°C and oil quenched. Tempering at 200–250°C results in a hardness of approximately R.C. 60. It is fairly brittle but reasonably easy to grind although it cannot be used much above room temperature since its strength falls rapidly with increasing temperature. It is comparatively inexpensive and has little distortion during hardening.

Double Six super die steel—This is hardened at 1,000°C by oil quenching after preheating to 800°C. Tempering at 200–400°C for two or three hours results in a hardness of R.C. 60. It is used to 500°C without loss of strength but grinding is more difficult than with K9 tool steel.

Stag Major high speed steel—This is similar to Double Six but can be used up to 650°C. It is hardened by preheating to 900–950°C and quenched by an air blast from 1,320–1,350°C. Tempering at 200–400°C gives a final hardness of about R.C. 60.

Tool steels are used for cylinders, anvils and pistons with support from more ductile steels or as loading pads where ductility is less important.

Maraging Steels

These contain much higher percentages of nickel and other metals such as molybdenum and cobalt and very low carbon content. They have the advantage of a comparatively simple hardening process which has only one stage and requires no sudden quenching. In addition the hardness does not depend strongly on carbon content as in the case of the low nickel alloy steels which leads to much less distortion. Maraging steels are fairly expensive and have a combination of high strength and high ductility under normal conditions, which are maintained at temperatures above and below ambient.

Stainless Steels

The steels described above are only suitable for operation in comparatively normal conditions, that is room temperature and absence of chemical attack. A series of high alloy content steels have been developed for high and low-temperature work and for maximum resistance to corrosion. They are known under the general heading of stainless steels and there are three types, martensitic, ferritic, and austenitic containing varying amounts of chromium, nickel and other elements. The ones normally used are the austenitic stainless steels with approximately 18 per cent chromium and 8 per cent nickel (*Table 1*, Appendix A). They are non-magnetic and can be hardened by cold working. In addition they retain their strength over a wide temperature range and are not attacked or embrittled by hydrogen, although under certain high-temperature

21

conditions they are attacked by nitrogen. An account of the use of austenitic stainless at low temperature is given in Chapter 7[5].

There are many special steels available which are comparable and may even be superior in high pressure properties to those given here and the reader is referred to manufacturers for further information (Appendix B).

Non-ferrous Alloys

Non-ferrous alloys used in high pressure apparatus include aluminium copper, aluminium zinc and beryllium copper. These are hardened by precipitation techniques. The normal procedure is to manufacture the alloy and rapidly cool it which leaves it soft and machinable. It is subsequently heated to a given temperature and held for some time during which precipitation of the alloying elements forms supersaturated solutions. The use of beryllium copper alloys is described in Chapter 7.

As in the case of steels, manufacturers provide details of heat treatment for particular alloys which are normally supplied in the soft condition, although many are prepared to supply to a given specification.

Tungsten Carbide

Cemented tungsten carbide is one of the hardest materials obtainable in large sizes and is used extensively in apparatus for very high pressures. In comparison with steels it has low ductility.

In the manufacturing process tungsten carbide powder is prepared from its elements and after thorough mixing with cobalt in varying proportions (from 3 to 30 per cent usually) it is hot pressed into required forms.

Cemented carbide is characterized by a high compressive strength (up to 50 kb) which is two or three times its tensile strength. Young's modulus is two or three times that of steel and therefore there is only a small amount of elastic deformation under load. The hardness is a function of the percentage of cobalt binder, the lower the percentage the greater the hardness; for example, on the Rockwell C scale 6 per cent cobalt samples are about 80 and 16 per cent about 68. A special advantage is that the hardness falls only 12–15 per cent on raising the temperature to 760°C where most steels have become extremely soft.

The main use is for pistons and anvils and loading pads where stresses are mainly compressive, but it is also used for cylinders with steel supporting rings. It is obtainable in sizes up to a few inches but larger sizes are more liable to contain flaws.

REFERENCES

[1] Foster, P. Field. *The Mechanical Testing of Metals and Alloys*. Pitman, London, 1948.

[2] *British Standard Mechanical Tests for Metals*. B.S. Handbook No. 13. British Standards Institution, London, 1951.

[3] Comings, E. W. *High Pressure Technology*. McGraw-Hill, New York, 1956.

[4] *Steel Specifications*. Issued by English Steel Corporation Ltd., Sheffield, 1963.

[5] Parker, C. M. and Sullivan, J. W. W. *Ind. Eng. Chem.* 1963, **55**, No. 5.

3

HYDROSTATIC PRESSURE APPARATUS

This chapter is concerned with apparatus which can be used for determining physical parameters up to pressures normally reached in truly hydrostatic conditions, the upper limit being approximately 30 kb at room temperature. Apparatus for pressures greater than 30 kb will be considered in subsequent chapters. Detailed accounts of techniques in the hydrostatic region can be found in the books by Bridgman[1] and by Hamann[2] for laboratory use and by Comings[3] for industrial scale applications. Here a description will be given of the fundamental properties of cylinders and pistons and of sealing and methods for entering probes in the general sense which includes many recent innovations not included in these books.

The piston and cylinder configuration forms one of the simplest methods of producing high pressures. The ultimate pressures which can be obtained, ignoring for the moment sealing and probe entry problems, is limited by the bursting strength of the cylinder and the crushing strength of the piston. The materials used in the construction depend on a number of practical factors and each specialist use requires individual treatment.

SIMPLE CYLINDERS

These are referred as 'simple' in order to differentiate them from compound types with multi-binding rings (Chapter 6). The theory of the distribution of stress in a cylinder wall subjected to an internal pressure has received considerable attention but only the case where elastic conditions are maintained throughout is amenable to a comparatively simple formal solution.

Thin-walled Cylinders

A formal solution for thin-walled cylinders may be used to a first approximation where the ratio of the external to the internal diameter (k ratio) is less than 1·10. It is assumed that the stress is uniform in the cylinder wall and as shown in *Figure 3.1*, the tangential stress is given by

$$T = Pr_1/r_2 - r_1$$

24

and the condition where this will exceed the proof stress Y of the material of the cylinder is

$$\frac{Pr_1}{r_2 - r_1} \geqslant Y \text{ that is } P \geqslant Y(k-1)$$

The bursting point is reached by replacing Y by the ultimate tensile strength σ. Since in this approximation $(k-1)$ is less than 0.1 only quite low pressures can be contained (of the order of 1 kb). The limiting criterion has been taken for the tangential stress since for a closed cylinder which is long compared with its radius the axial stress is a factor of two less than the tangential value.

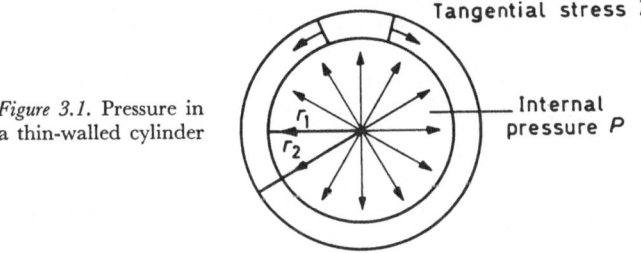

Figure 3.1. Pressure in a thin-walled cylinder

Tangential stress T

Internal pressure P

Thick-walled Cylinders

These may be defined as having k ratios larger than 1.10 which will be the case for most practical purposes. The essential difference from the thin-walled case is that the assumption of uniform stress throughout the wall no longer holds. The stress due to an internal pressure can be analysed by the Lamé–Clapeyron equations[4] if the whole wall remains inside the elastic region of the constituent material. The yielding criterion is more difficult to define and generally speaking two conditions for exceeding the elastic limit are considered. These are the elastic breakdown pressure at the bore and the overstrain pressure at the outside surface. Thus, as pressure increases a point is reached where the inner layer is stressed beyond the elastic limit and a boundary between plastic and elastic material gradually moves through the wall until it reaches the outer surface at the overstrain pressure.

The elastic breakdown pressure P_e is not simply the tensile yield stress since the conditions under which the latter is defined differ from those in a cylinder wall where a small elemental volume is subjected to axial, radial and tangential stresses. There are several

25

calculated criteria for estimating the elastic breakdown pressure and the one which is normally used because it is confirmed by experiment reasonably well is the Von Mises or octahedral-shearing stress formula[3]

$$P_e = \frac{Y}{\sqrt{3}} \frac{k^2 - 1}{k^2} \text{ (for a closed ended cylinder)}$$

The formula predicts that for k greater than 4, P_e approaches a limiting value of $Y/\sqrt{3}$. However the overstrain pressure limit can be shown to be proportional to the logaritham of k which means that a cylinder can be used to much higher pressures than P_e before a completely plastic state is reached[3]. For example a cylinder ($k > 3$) of EN 26 steel hardened to a value of Y of about 12 kb will withstand 7 kb internal pressure before any plastic distortion at the bore.

Beyond the elastic breakdown limit the cylinder bore eventually bursts under pressure. The conditions for this have been calculated. The Lamé criterion is

$$P_B = \sigma \left(\frac{k^2 - 1}{k^2 + 1} \right) \text{ (σ is the ultimate tensile strength)}$$

A more well-known formula which has been tested with steel cylinders of k values up to 7 is the mean diameter formula[5]

$$P_B = 2\sigma \left(\frac{k - 1}{k + 1} \right)$$

The latter is the more normally used criterion.

More formal approaches to the problems of inelastic deformation in thick cylinders requires considerable experimentation and calculation and will not be considered here. The reader is referred to the papers by Manning[5] for further details.

Autofrettage

The above discussion has ignored any residual stresses in a cylinder which has been subjected to heat treatment and cold working which can extend the elastic limit by up to 50 per cent in some cases. The process of autofrettage has been used to advantage by many high-pressure experimenters. In this, pressure is generated in the bore of a cylinder which is in excess of the elastic breakdown pressure and hence causes the material to yield. The pressure is removed and a stress-relieving treatment applied to restore stable conditions but not to such an extent that the benefits of the cold working are removed. After a number of cycles to stabilize the cylinder a considerable increase in the yield stress of the bore is obtained. Essentially autofrettage treatment leaves compressive stresses in the material which have first to be overcome with increasing pressure in the bore.

External Pressure

The usable limit of a cylinder is increased by applying external pressures.

(i) By shrunk-on or press-fitted supporting rings. This is discussed in detail in Chapter 6.

(ii) By the application of hydrostatic pressure. Up to 100 kb have been contained in cylinders with this method of external support but since the apparatus becomes complicated it is not used very much.

(iii) From the support by gradually forcing the cylinder into a supporting block. This is discussed on page 48 and allows pressures of up to 30 kb to be generated in the bore.

Lapping and Honing

For the best results a cylinder bore should be honed. In this process a special honing tool, of hardened steel usually, is used to smooth the bore after grinding. This results in a very smooth and accurately symmetrical bore with a considerably work hardened surface which improves the performance and sealing under pressure.

The surfaces of a piston should be polished or lapped to a finish better than 2×10^{-5} cm with diamond paste. This again results in advantageous surface work hardening and good sealing qualities.

Summary

The upper pressure limit of an untreated cylinder ($k > 3$) within the elastic limit of the bore regions is equal to $Y/\sqrt{3}$ which means about 10–12 kb for the hardest ductile steels. This limit can be increased by as much as 50 per cent by autofrettage and by factors of up to three by the application of external pressure. Cylinders may be taken up to much greater pressures than the elastic breakdown value before all the material is in a plastic state.

The materials employed in cylindrical pressure vessels include alloy, stainless and tool steels, pure beryllium, beryllium–copper alloys, tungsten carbide, sintered alumina, aluminium alloys, sapphire and diamond, and which of these is to be used depends

27

obviously on the experiment in mind. For example, diamond and beryllium are fairly transparent to x-rays; sapphire and diamond transmit a wide range of visible and infra-red radiation; beryllium–copper is non-magnetic; sintered alumina will withstand 1,000°C at 50 kb if suitably supported; some aluminium alloys are comparatively transparent to neutrons etc. The use of many of these materials is described in subsequent chapters.

PISTONS

The ultimate strength of pistons is much simpler to define since the operative factor is the compressive strength of the particular material. Some loss of ductility is exchanged for increased hardness and tool steels may be use up to 25–30 kb force over area and tungsten carbide, which has probably the greatest strength, up to approximately 50 kb. Piston life is impaired by large out-of-balance forces and therefore their lengths should be as short as possible. Conventional pistons can be used up to much higher pressures by subjecting them to external support as in the case of cylinders and in the next chapter an account is given of the principle of massive support through which usable piston strengths are increased to 500 kb.

SEALING

Figure 3.2 illustrates the most basic type of general static sealing where the ultimate pressure contained is the limiting force which can be applied to the gasket material by tightening the bolts. This is reached when the pressure medium pushes out the gasket material.

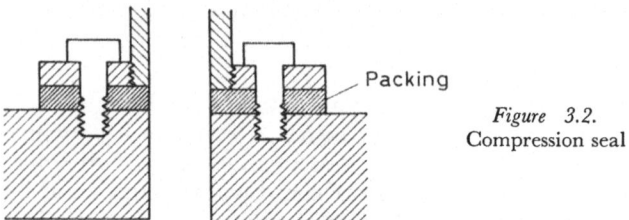

Packing

Figure 3.2.
Compression seal

Modifications to prevent extrusion of the gasket material have been used but since they depend on the application of external forces which are larger than that to be contained they are not completely satisfactory. Two other versions of the external force seal are shown in *Figure 3.3*, where the applied force produces a high pressure over

28

a small area along the line of contact and hence makes a satisfactory seal. These may be used to several kilobars.

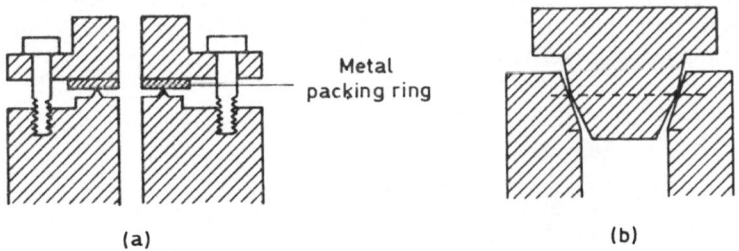

|(a)| |(b)|

Figure 3.3. (a) Ipatieff knife edge closure for use up to a few kilobars. (b) Line-contact seal with cone angles 59° and 60° for use up to a few kilobars.

The pressure range is extended to tens of kilobars using an alternative method in which the pressure in the fluid is used to make the seal. Various forms are shown in *Figures 3.4* and *3.5*.

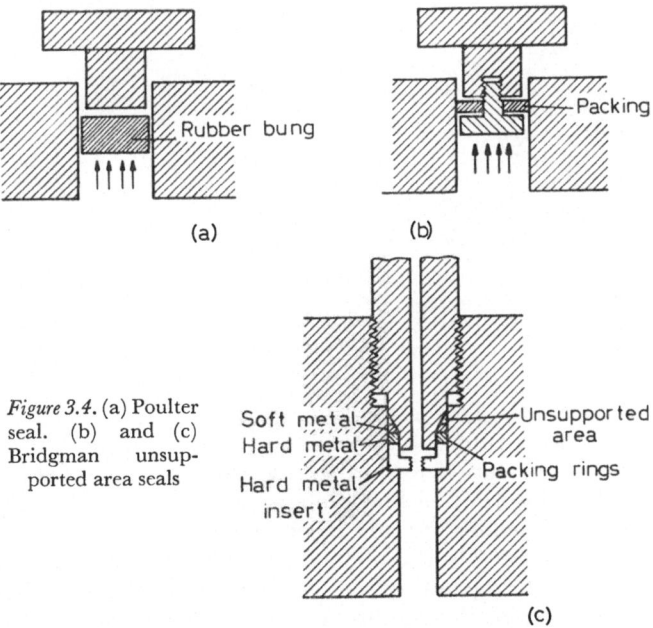

Figure 3.4. (a) Poulter seal. (b) and (c) Bridgman unsupported area seals

There are two prime effects when considering pressure generation in a fluid with a piston and a cylinder. These are leakage of the fluid and friction of the piston in the bore and unfortunately they are mutually incompatible since the looser the fit of the piston the greater the leakage. In practice leakage is many times more important than friction since the latter can be overcome in most instances by increasing the applied load.

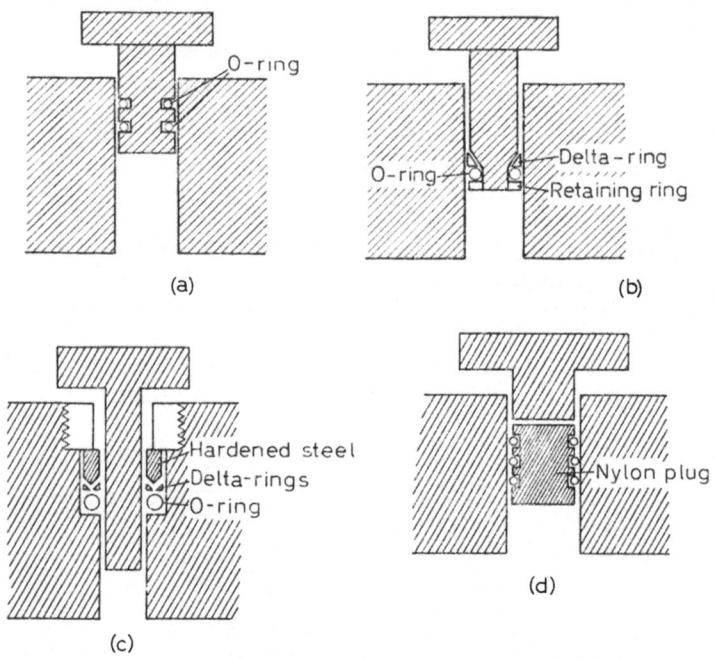

Figure 3.5. (a) O-ring seal. (b) O-ring and delta-ring seal. (c) Patterson seal. (d) Nylon plug seal

The Poulter seal—This is the simplest type of self-sealing closure and is basically a rubber bung which fits closely in the bore of the cylinder. The pressure deforms the bung laterally and makes a seal. It will hold pressures of a few kilobars but it is difficult to remove and cannot be used for pressure media of low viscosity (*Figure 3.4a*).

The Bridgman seal—In this method the unsupported area principle first defined by Bridgman is used. As shown in *Figure 3.4b*, the total pressure exerted on the mushroom head is transferred to the soft packing which has an area less than that of the bore since the stem is floating. This means that the actual pressure in the packing is higher by a ratio which is simply the total area to the unsupported

30

area. Thus no matter what pressure is to be contained the pressure in the seal is always higher. For pressures up to 30 kb and normal temperatures the packing usually consists of a sandwich of thin rubber between two annuli of annealed copper or mild steel. For high temperatures indium or lead is substituted for rubber. Friction is often as high as 10 per cent of the load but it can be reduced by minimizing the thickness of the packing. The stem of the mushroom head usually has a screw thread so that a suitably tapped rod can be used to remove the seal after withdrawal of the piston. The main disadvantage of the seal is that if too great a force is produced in the packing the stem may be snapped off and/or the cylinder wall compressed beyond its elastic range in the sealing region. As a general rule the stem diameter should be about half the total bore diameter.

Figure 3.4c shows a second version of this method of sealing which is used for end plugs rather than moving pistons. Under pressure the soft metal ring deforms against the chamfered edge of the hard plug and the unsupported area is that part not in contact. As in the first case a sandwich of different materials is usually used to make satisfactory seals at low pressures although at higher pressures a single ring of mild steel is satisfactory. An initial seal is made by tightening against an insert which also serves as a method for withdrawing. The rings have to be renewed when after repeated use there is no further undeformed area.

O-ring seals—O-ring seals are quite simple in construction and operation. The material may be neoprene, nylon, or similar substances. They are fitted into grooves in the piston or in the walls of the cylinder as shown in *Figure 3.5a*. For the highest pressures two or three O-rings are needed and it would appear that the leakage of some fluid past the first or second O-ring allows pressure to be set up on either side preventing further leakage. This type of seal has the advantage whereby the whole of it is dissembled with the withdrawal of the piston. As a general guide the depth of the groove should be about 0·8 of the thickness of the O-ring (although values between 0·7 and 0·95 have been quoted by various users) and optimum radial clearance of the piston is about 0·005 mm in a 10 mm diameter bore. A disadvantage which is usually not too serious is the weakening of the pistons or cylinder walls by machining grooves in them.

Several modifications of the O-ring seal have been employed at the highest pressures. One drawback with the simple type just described is that although the friction is initially only a few per cent,

31

it rises rapidly with increasing pressure cycling due to extrusion of the rubber into the clearance region. In the version used at the National Engineering Laboratory and by the Harwood Engineering Company (*Figure 3.5b*) a ring of triangular cross section (delta ring) of phosphor bronze or beryllium copper is used to reduce extrusion, and combined with a single O-ring pressures up to 10 kb can be contained with as low as 1 or 2 per cent friction at the seal. It is most useful for static closure plugs but when used for moving piston seals above 10 kb the delta rings eventually extrude and since this is due to the increase in clearance with increasing pressure harder materials used for the rings are not an improvement generally.

A simple extension of the O-ring and delta ring seal is the Patterson type[7,8] in which the piston clearance at the moving seal is not changed appreciably with increasing pressure (*Figure 3.5c*). The advantage here is that the seal is static and accurate fitting of the delta rings is unimportant. N. B. Owen, at the National Physical Laboratory, has used double delta rings made in one piece of phosphor bronze (or nylon) which have proved satisfactory to 10 kb with friction of the order of 3 per cent.

A piston seal using a separate nylon plug is shown in *Figure 3.5d*. The O-rings make a satisfactory seal up to 10 kb and use of the plug avoids any special machining of either the piston or the cylinder wall but removal is sometimes difficult.

In conclusion the O-ring seal is a very simple one to make, especially with the availability of a wide variety of shapes and materials, and to use successfully to 10 or 12 kb with little friction. Hollow metal O-rings (nickel) coated with teflon have been used with beryllium–copper delta rings to 20 kb by S. L. S. Thomas at the National Physical Laboratory. The ring is perforated and fluid inside it helps to make the seal. Neoprene can be used up to 190°C without any serious deterioration and nitrite rubber up to 230°. Above this metal types are employed.

The Bridgman seals contain greatest pressures if properly made but friction tends to be much higher. Both types will seal liquids of widely different viscosity.

PERMANENT CLOSURES

These are generally not very useful for laboratory apparatus since they make frequent cleaning and the removal of samples from the inside of a cylinder a comparatively long and difficult operation.

They are more common in industrial process apparatus and a comprehensive account has been given by Comings[3].

HIGH-PRESSURE PIPING CONNECTIONS AND VALVES

In many cases it is necessary to connect high-pressure apparatus to a fluid compressor by means of piping and valves. There are numerous types available and the reader can gain further information by consulting the catalogues of some of the firms listed in Appendix B. High-pressure tubing is normally made in standard sizes from alloy steels or similar materials. The ratios of outer to inner diameters vary between 3 and 5 and usually the smallest tubes carry the highest pressures. In some cases the tubing is lined with copper or some other substance to reduce corrosion by particular fluids.

There are a variety of methods for connecting high-pressure tubing and only two examples are given here (*Figure 3.6*) (for others

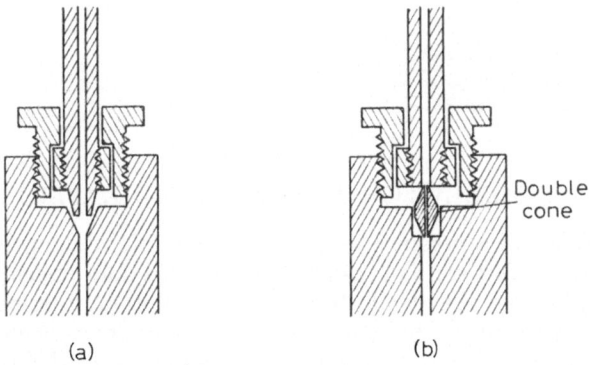

(a)   (b)

Figure 3.6. (a) Line-contact piping seal. (b) Double cone seal

see Comings[3]). The most common type employs a cone ending with 59° included angle which makes a line seal against a 60° included angle cone insert in the main body of the apparatus. The screw thread on the end of the pipe can be eliminated by welding on a cone end of slightly larger size. This type of seal is effective to 10 kb for tubes less than 0·5 cm internal diameter. Another type is shown in *Figure 3.6b* which uses a double cone of hardened steel to make a line of contact seal between the connecting units and is usable to at least 10 kb.

As in the case of piping connections there are many designs for high-pressure valves and as these are also readily available commercially there is little point in giving much detail in this book. *Figure 3.7* illustrates a simple valve with a non-rotating stem and a Bridgman unsupported area seal. The valve seating is a steel ball or a cone making a line of contact seal.

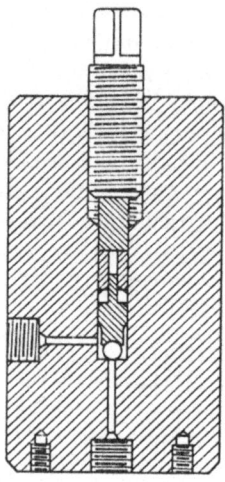

Figure 3.7. High-pressure valve (Harwood Engineering Co.)

PRESSURE MEDIA FOR HYDROSTATIC EXPERIMENTS

The choice of a pressure medium is inevitably a compromise of a number of factors. For example, electrical and thermal conductivity, transparency to electromagnetic radiation, freezing point, viscosity, chemical reactivity and corrosivity and depends ultimately on the experiment in mind. The most commonly used liquids are the lower alcohols, kerosene, pentane, iso-pentane, and light mineral oils and some of their relevant properties are listed in *Table 3.1*. These are all good electrical insulators and have negligible chemical effects on normal solids. According to Bridgman[9] isopentane is liquid up to 30 kb at room temperature although Reeves *et al.*[10] have found the limit to be nearer 20 kb. Hamann[2] has used some alcohols to 40 kb at room temperature due to the occurrence of a supercondensed phase. The choice of a liquid medium must rest to a considerable extent not only on viscosity and solidification pressure but also on the volume change during compression since this may be a limiting factor on the size of a sample and the high

34

pressure container. Glycerine has a low compressibility but becomes extremely viscous above 4 kb. Carbon disulphide is used as a pressure medium in neutron diffraction experiments since neither carbon nor sulphur are strong absorbers of thermal neutrons. Its freezing point is 12 kb at room temperature and at −50°C at atmospheric pressure, but it is found that in most cases if pressure is applied in the liquid state fairly isotropic conditions are maintained through the freezing point. This behaviour is observed also in the case of helium and hydrogen when used as solid pressure media. The reader is advised to consult Bridgman[9], Hamann[2] and Reeves et al.[10] for further details of liquids under compression.

Table 3.1

(After Reeves et al.[10], and Hamann[2])

Substance	Melting pressure at 25°C kb	Viscosity at 25°C η_{20kb}/η_0
^{4}He	120	
H_2	55	
N_2	24	
Ar	14	
CH_3OH	20	50
n. C_5H_{12}	15	1000
i. C_5H_{12}	21	1000
Shell Tellus Mineral oil	~ 10–12	
CS_2	12	

As a rule gases provide less problems than liquids as pressure media. Of the commonly used gases, hydrogen and nitrogen can cause embrittlement of certain alloy steels but not stainless steel, and helium tends to penetrate otherwise non-porous materials. Argon is a very suitable gas for use at room temperatures and above. The respective solidification pressures are shown in *Table 3.1*.

In conclusion, for straightforward experiments in which the only requirements are high electrical insulation, high freezing pressures at normal temperatures and a minimum of chemical corrosivity, the most useful pressure medium is an alcohol or isopentane, which is obtained easily from suppliers of chemical reagents.

LOAD GENERATION

The load required to produce a given pressure in a cylindrical vessel depends obviously on the size of the bore. There are practically three basic methods for applying such forces.

Large Scale Presses

Most high-pressure laboratories have large presses of the type listed briefly in Appendix B with capacities of the order of 200 or 300 tons or more. A piston and cylinder device may be placed in these and the required pressure generated. On a smaller scale there are available presses of up to 30 tons capacity which are very convenient for the more usual size of laboratory equipment. Essentially this method of force production involves bringing high-pressure apparatus to the load-producing equipment.

Liquid and Gas Compressors

Liquid compressors up to 5 kb are readily available in the form of small hand or motor driven pumps. Gas compressors tend to be more bulky and the reader is referred to makers for more details.

Hydraulic Intensifiers

This is a method of load production which was used extensively by Bridgman. Generally speaking the press is incorporated into the high-pressure apparatus. Comparatively low hydrostatic pressure is applied to a piston of large area which is connected directly to a much smaller piston (*Figure 3.8*). The pressure exerted by the second

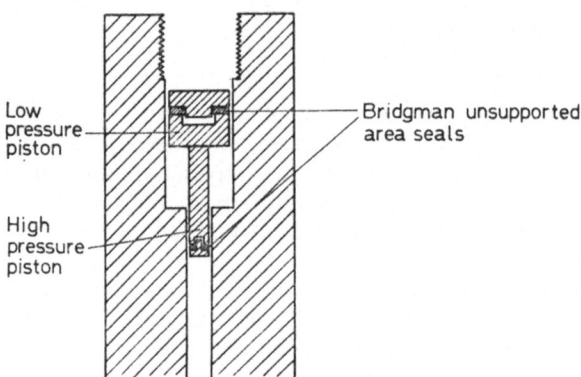

Figure 3.8. Hydraulic intensifier

piston is the original pressure multiplied by the ratio of the areas of the two pistons. In practice this factor can be at least 40 or 50 and hence pressures of a few hundred bars can be used to produce tens of kilobars. This is a very convenient method since small hand pumps capable of supplying oil pressures of 1·5 kb are readily available. An added advantage is that using high-pressure tubing the experimental apparatus can be placed some distance physically from the pressure generator, which is of obvious importance when making x-ray and neutron diffraction measurements. Intensifiers can be very compact and have been used to produce pressures up to 500 kb (see Chapter 4).

The details of an actual intensifier are described in connection with a hydrostatic apparatus on page 44.

MEASUREMENT OF HYDROSTATIC PRESSURES

The most fundamental measurement is made by balancing against a mercury column but this is limited to comparatively low pressures and the normally used primary gauge is the dead weight or free piston gauge. There are several versions but only a simple form will be described here (*Figure 3.9*). A known pressure is

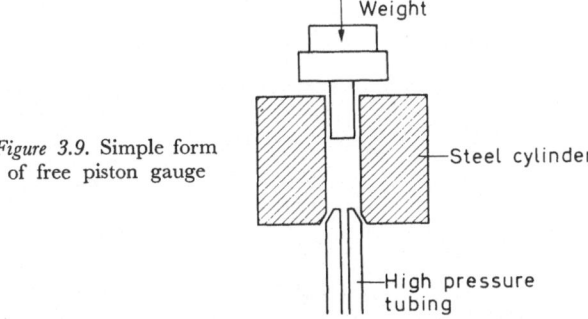

Figure 3.9. Simple form of free piston gauge

generated in the oil by placing a weight on the piston. Usually friction is reduced by rotating the piston and allowing a very small leakage which is made up from an injection valve. The pressure range is altered for a particular gauge by using cylinders and pistons of different diameters. Accuracies are very high after appropriate corrections.

The area of the piston is either measured directly or determined from a low-pressure comparison with a column of mercury. The main error is due to elastic distortion of the piston and cylinder at higher pressure under a combination of axial and radial stresses. In the method of Dadson[11] an effective area A is defined, where A_0 is the measured or low-pressure value

$$A = A_0 \left(1 + \alpha F(P)\right)$$

37

$F(P)$ is a complicated and not easily determined function of pressure and α is inversely proportional to the elastic modulus. If two such devices of similar size but of materials differing considerably in elastic modulus are compared by balancing pressure against each other, then A can be determined.

For a hardened steel piston and cylinder the effective area changes by about 1 part in 2,000 per kb. The capability of the method is demonstrated by Dadson and Greig's[12] determination of the freezing pressure of pure mercury at 0°C as $7\cdot569 \pm 0\cdot001$ kb (*Figure 3.10* illustrates the experimental arrangement).

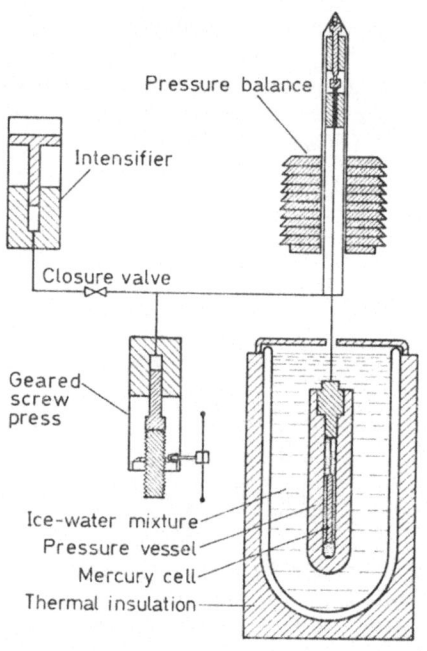

Figure 3.10. Experimental arrangement for determining the freezing pressure of mercury at 0°C (after Dadson and Greig[12])

Although the free piston gauge is absolute in the broadest sense it is nevertheless rather complicated for normal application in high-pressure experiments and secondary methods described below are usually employed. Recently Heydemann[22] has used a dead weight piston gauge to determine the bismuth I—II transition.

Bourdon Tube

This is the most widely used gauge for pressure less than 3 kb. It is calibrated against free piston gauges. The pressure unit is a steel tube of ellipsoidal cross-section bent into an arc. One end is closed off and pressure applied to the inside tends to straighten the tube. The movement is transferred to a mechanical lever system to give a dial reading as shown in *Figure 3.11.* The range of

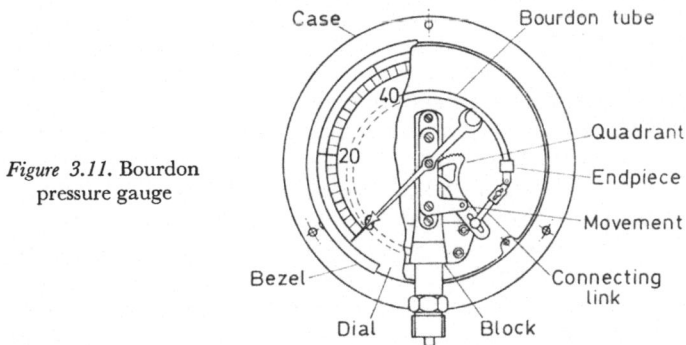

Figure 3.11. Bourdon pressure gauge

Budenberg pressure gauges uses this principle. Accuracies are about 0·1 per cent of the maximum scale reading for the best types.

Electrical Resistance Gauges

Many experimenters have found that the best continuous pressure scale up to 25 kb is that indicated by the electrical resistance of manganin wire coils. It is well known that the resistance of manganin increases linearly with pressure up to 12 kb and has a deviation of only 2 per cent of this linearity at 25 kb. Manganin wire is usually 80–84 per cent copper, 4–5 per cent manganese, 12 per cent nickel and the remainder iron, and a convenient gauge may be made by winding non-inductive coils of about 100 ohms resistance from 40 gauge silk covered wire. Comparisons with a free piston gauge have shown that the pressure coefficient which is of the order of $2·4 \times 10^{-4}$ bar^{-1} is linear to within a few tenths of 1 per cent up to 12 kb and therefore for most purposes a single calibration point is adequate. This is usually the mercury freezing point at 0°C (7·569 kb). At the upper end of the scale the bismuth I–II transition at 25·4 ± 0·2 kb is used. The pressure coefficient of resistance depends on a number of factors and its value for two samples cut from the same wire spool can differ by up to one or two per cent. For optimum performance the coils are put through the Bridgman cycle in which they are exposed to temperatures of about −75°C for two hours and then 120°C for eight hours. This is repeated several times with simultaneous applications of pressure over the operative region. It is found that maximum reproducibility is obtained only with hard soldered joints. Above room temperature manganin has a comparatively large temperature coefficient varying between $+1 \times 10^{-5}$ and -8×10^{-5} deg^{-1} over the range 5° to 100°C and when it is recalled that increasing pressure heats a medium considerable care must be taken to allow for a return to equilibrium before measurements are made. Darling and Newhall[13] proposed a resistance gauge using an alloy of gold with 2·1 per cent chromium which has a much lower temperature coefficient of resistance particularly between room temperature and 100°C (less than 4×10^{-6} deg C^{-1}). Its pressure coefficient of resistance is only about one third of that for manganin but this is not a major deterrent for most purposes. Gold–chrome gauges have the advantage of requiring less heat and pressure seasoning and Darling and Newhall claim that a single pressure cycle is sufficient to give consistent measurements, although other experimenters have encountered more difficulty[14]. A typical coil would consist of 36 gauge wire wound non-inductively and stabilized by baking at 150°C for 36 hours. The disadvantage with gold–chrome gauges is that solder reacts with the wire making spot welding essential.

To summarize, the most convenient method of measuring pressure is probably with a manganin wire gauge. The scale is linear with a deviation of less than 0·5 per cent up to 12 kb and less than 2 per cent up to 25 kb. There are a number of phase transitions which can be detected volumetrically and used as extra calibration points between 0 and 25 kb and at the top end of the scale there is the bismuth I–II transition at 25·4 kb, which has been determined accurately by Kennedy and La Mori[15] (Chapter 1).

ELECTRICAL CONNECTIONS

Although several different types of seal for electrical connections have been used, the most useful one for the highest pressures appears to be that due originally to Amagat and which essentially invokes the unsupported area principle as shown in *Figure 3.12*. It is essential that the steel cone and ceramic insulator are lapped into the hole in the end plug otherwise there will not be a good seal at low pressures. (This is not so important if unfired pyrophyllite is used instead of the ceramic.) The seal can be used up to 30 kb at well above room temperatures. A disadvantage is the degree of capacitance between the inner and outer conductors and only one connection can be brought out in each unit.

A modification which is usable to 8 kb and is relatively simple to make is that suggested by Gugan[16] and used extensively at the

National Physical Laboratory by the author. As shown in *Figure 3.13* a conical hole is used as before at the high-pressure interface. This terminates in a comparatively long lead hole filled by a close-fitting multi-bore ceramic tube, for example Degussite tubing with four holes 0·25 mm diameter in a total diameter of 1·5 mm.

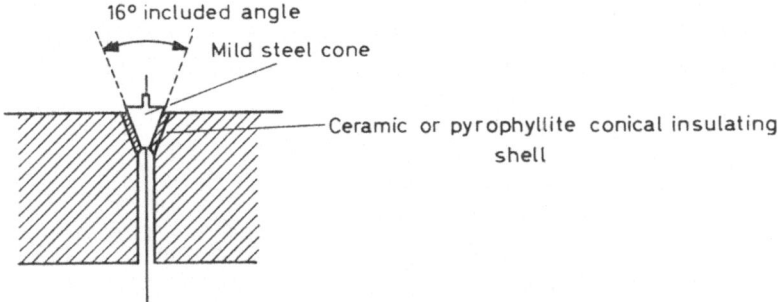

Figure 3.12. Amagat electrode seal

Wires of appropriate gauge are sealed into this tube with epoxy resin. The conical hole is filled with epoxy and any air bubbles removed by warming gently with a hair drier. After setting the seal is heated to about 150°C for half an hour in order to complete the hardening process. The grade of epoxy resin used will depend on the temperature range to which the seal is to be subjected. To remove the seal

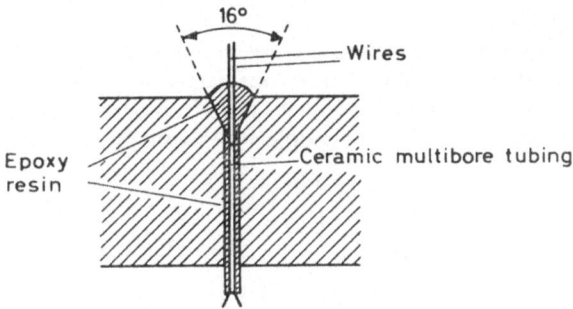

Figure 3.13. Electrode seal after Gugan[16]

the whole unit should be heated to the point when the epoxy resin carbonizes and the holes are then easily cleaned out. This latter operation should not be carried out unless it is known that no detrimental effect can be caused to the steel.

41

This electrode seal is suitable to 8 kb and temperatures below 150°C. It is important that there is always adequate safety shielding since any failure results in a fine jet of liquid through one of the bores which is a very dangerous hazard. The epoxy resin type of seal is an especially convenient one for getting thermocouples into a high-pressure apparatus since there is no intermediate connection to metal cones.

A combination of the metal cone insert and epoxy resin plug is used in the seal of Blosser and Young[17]. As shown in *Figure 3.14* a

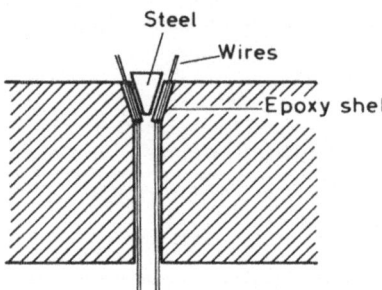

Figure 3.14. Electrode seal after Blosser and Young[17]

very large number of wires can be introduced through a single seal. The hardened steel cone insert (16° included angle) acts as a structural component only and since there is no brazing or soldering to it heat distortion leading to poor sealing qualities is eliminated. The wires are fixed into the conical epoxy resin shell in a separate mould and the seal assembled in the same manner as the Amagat type. Continuous thermocouple wires can be introduced and pressures up to 10 kb contained in either a gas or liquid.

Another seal for comparatively high temperatures (200°C) is shown in *Figure 3.15*. Thermocouple wires embedded in magnesium

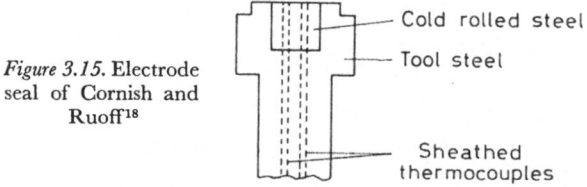

Figure 3.15. Electrode seal of Cornish and Ruoff[18]

oxide in a stainless steel sheath (Philips coax.) are silver soldered into a steel insert which is then shrunk-fitted into the main body of a plug. This holds liquids up to 10 kb but gases tend to leak through the magnesium oxide; this can be prevented with epoxy resin.

WINDOWS

At low pressures, that is less than 1 kb, a conical window can be constructed in a similar form to that for the electrode connection shown in *Figure 3.12* using a glass instead of steel cone insert. Unfortunately, on unloading friction holds the window in place and frequently results in its shattering. Nevertheless windows of cone-shaped sapphire crystals lapped directly into apertures without any liner have been used to 10 kb for microwave transmission into high-pressure apparatus (see Chapter 7).

For pressures up to 10 or 15 kb the most useful design for a radiation window in the spectral range from x-rays to the far infra-red is the Drickamer[19] design using a Poulter packing technique which is also an unsupported area application (*Figure 3.16a*). The

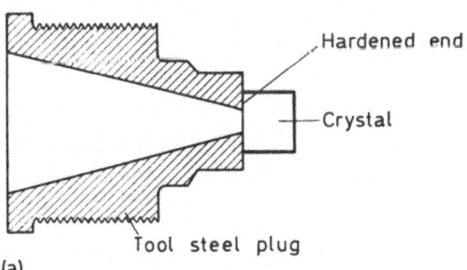

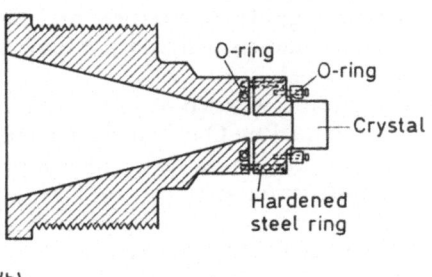

Figure 3.16. High-pressure window plugs (a) after Fishman and Drickamer[19] (b) modification by N. B. Owen (N. P. L.)

seal is made between the material of the window and a hardened steel plug giving an aperture which is the unsupported area. The two surfaces should be lapped to a finish of better than 0·00025 mm and are 'rung' together by sliding the two surfaces in contact under finger pressure with the exclusion of air. The 'ringing on' can be tested generally by seeing whether the plug, which might weigh more than 500 g for a 1 cm window, can be suspended by holding the window. Materials depend on the experiment in mind, for example synthetic sapphire has good transmission between the near ultra-violet and 4 microns wavelength and can withstand pressures up to 15 kb and more and is available in sizes up to 2·5 cm diameter. Crystal quartz is transparent in the far infra-red (50 to 1,000 microns) but is much weaker than sapphire. Diamond has good transmission over a wide range of frequencies. It is also the strongest single crystal material but is only available conveniently in comparatively small sizes (a 2·5 mm diameter window might cost £50). The relative transmissions of a number of window materials are shown in *Figure 5.24* (Chapter 5).

An alternative method of making the seal is to use O-rings as shown in *Figure 3.16b*. This is often more satisfactory at low pressures with fairly low viscosity fluids.

It is desirable to have the length-to-diameter and supported-to-unsupported-area ratios as high as possible because of large bending stresses set up under pressure.

APPARATUS FOR ELECTRICAL RESISTANCE, OPTICAL ABSORPTION AND COMPRESSIBILITY MEASUREMENTS TO 12 KB

In the preceding sections a general account has been given of the techniques used to contain pressures in fluid media. In the following a complete apparatus used for studying physical properties of solids and liquids is described, and the incorporation of many of the above techniques will afford a basis on which similar apparatus can be designed.

The apparatus is based on one used by Fishman and Drickamer[19] and was designed by Owen at the National Physical Laboratory for electrical, optical absorption and compressibility measurements.

The construction is shown in *Figures 3.17* and *3.18*. The main high-pressure cylinder is a block of EN 30 B steel hardened and tempered to R.C. 48–50 bored out longitudinally with a 1·27 cm diameter hole and three other apertures with profiles as shown. All the regions where seals are to be made are honed. The upper part of the cylinder is elongated to connect with the intensifier. The high pressure piston is made of K 9 tool steel (R.C. 58–60) and is lapped into the bore to give a clearance of less than 0·0075 mm. The piston is sealed with the Bridgman arrangement using a tool

44

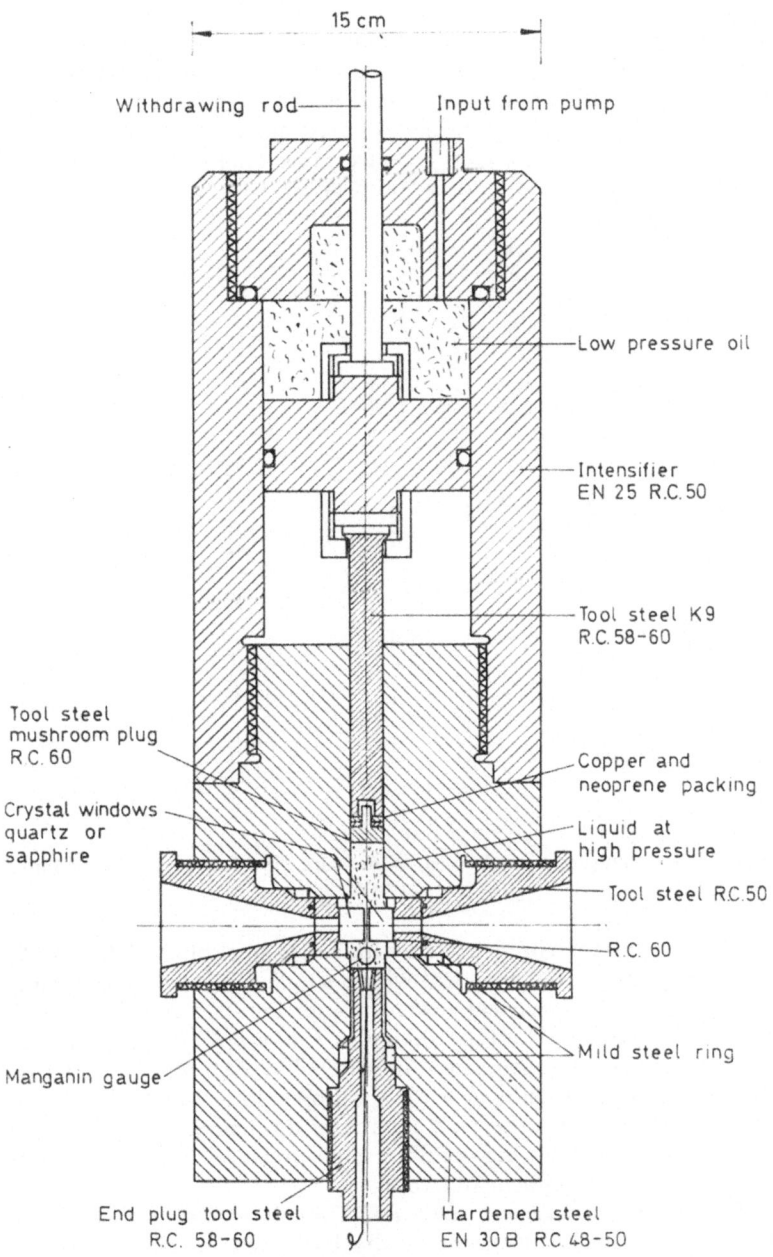

15 cm

Withdrawing rod

Input from pump

Low pressure oil

Intensifier
EN 25 R.C. 50

Tool steel K9
R.C. 58-60

Tool steel
mushroom plug
R.C. 60

Copper and
neoprene packing

Crystal windows
quartz or
sapphire

Liquid at
high pressure

Tool steel R.C. 50

R.C. 60

Mild steel ring

Manganin gauge

End plug tool steel
R.C. 58-60

Hardened steel
EN 30 B RC 48-50

Figure 3.17. 12 Kilobar optical high-pressure cell

steel mushroom head and a packing sandwich of rubber between two copper annuli. The friction of this is about 5 to 10 per cent for a total thickness of 3 mm. The end which is connected to the low pressure piston of the intensifier is flared out to twice the diameter in order to reduce contact pressure. The length depends on the required stroke and is made appropriately.

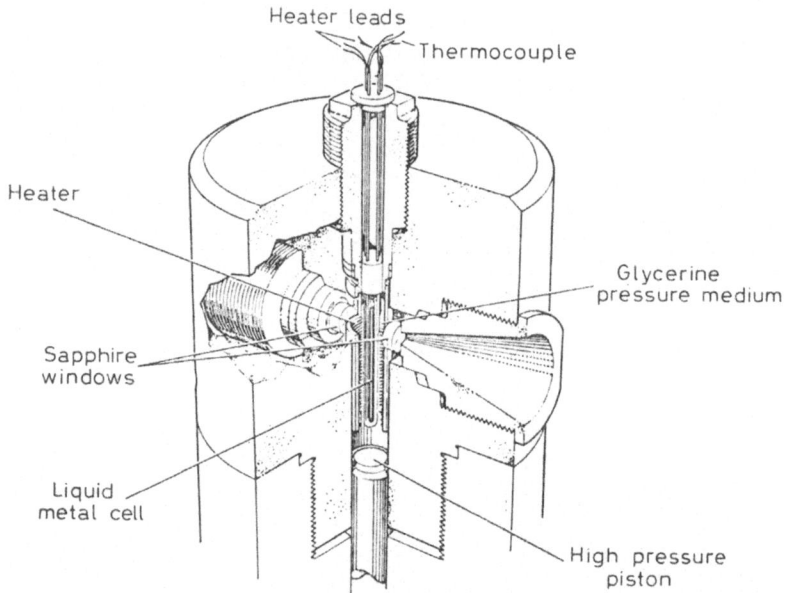

Figure 3.18. High-pressure cell used to measure electrical properties of liquid metals

The bottom aperture is closed with a tool steel plug (R.C. 58–60) using an unsupported area seal with a single soft steel packing ring. The plug carries electrical leads through four 2·5 mm holes with epoxy resin seals of the type shown in *Figure 3.13*. These electrode seals reduce the upper pressure limit to below 9 kb but with alternative methods or absence of leads the upper limit is 12–14 kb. The plug has flats on its outer end for initial tightening with a spanner against the packing ring, and its high-pressure end has a screw thread for attaching a heater tube.

The window plugs are made also from tool steel hardened and tempered to R.C. 50 except at the high-pressure end where the window seal is made. The hardness at this point is R.C. 58–60.

The window plugs are sealed into the cylinder in the same manner as the electrode carrying end plug. The window plugs are designed with an aperture of f.2 because one of the uses was for infra-red absorption experiments where sources are fairly weak requiring a maximum throughput of energy. The windows are single crystals of sapphire or quartz 1·27 cm diameter and 1·27 cm long and are 'rung on' to the especially hardened end of the plugs after lapping the appropriate surfaces to a finish of better than 0·00025 mm. A brass keeper is placed around the crystals to avoid accidental knocking while the plug is screwed into the cylinder. The gap between the two windows is altered by using sealing rings of different thickness but it is obviously fixed once a seal has been made. The window plugs also have flats on the outer ends for initial tightening. By replacing the window plug seals by O-rings the distance between the crystals can be varied while the apparatus is under pressure.

The intensifier is a conventional one made from EN 25 steel (R.C. 50). The area ratio of the low and high-pressure pistons is about 40. Low-pressure oil is supplied by a hand pump through a cone connection as shown, O-ring seals being employed throughout. The low-pressure piston is made of tool steel (R.C. 58–60) and is loosely locked to the high-pressure piston by a nut, so that the two are withdrawn together which would not otherwise be the case because of friction. For the same reason a rod is attached to the low pressure end as shown for full withdrawal after pressure has been released. The total height of the intensifier is about 30 cm and it is screwed onto the cylinder as shown.

Pressure is measured either by noting the oil pressure in the intensifier with a Bourdon gauge and using a friction connection or by a manganin resistance manometer directly in the high-pressure bore.

Electrical Measurements in Solid and Liquid Metals

Electrical resistance and thermo-electric measurements are made using the arrangement shown in *Figure 3.19*. For this blank window plugs are used and the whole unit is held so that the end plug is uppermost. An economy of leads is achieved by using the thermocouples as current and voltage probes for resistance measurements. A nichrome heater wound on a ceramic former is screwed onto the end of the plug. (An account of the effect of pressure on thermocouples is given in Chapter 1.) An efficiency of about 1° per watt is usually attainable and thermal gradients are reduced by tapping the winding and using variable shunt resistors. Pressure measurement is made with a manganin gauge calibrated at the freezing pressure of pure mercury contained in the glass tube.

Compressibility Measurements

These are made by observing a meniscus through sapphire windows.

Optical Absorption Measurements in Organic Liquids

For optical experiments in liquids the sample can be the pressure medium. In this case it is an advantage to have the window positions continuously adjustable for changing the sample thickness. The unsupported area steel ring seals are replaced by rubber O-rings with nylon double delta ring types. Sapphire windows are used for measurements in the wavelength region 0·2 to 4 microns and hold up to 12–14 kb.

For far infra-red (wavelengths greater than 50 microns) absorption experiments, the sapphires are replaced by quartz single crystals. These are weaker and the upper pressure limit is only 6 to 8 kb. The manganin resistance manometer is calibrated by noting optically phase changes for organic substances such as carbon tetrachloride (see Bridgman[9] for data on phase changes). Far infra-red spectroscopic techniques are described in more detail in Chapter 5.

It must be emphasized that a liquid under pressure has considerable stored energy and it is extremely dangerous to proceed without taking adequate shielding precautions, particularly in respect of window and electrode seals.

EXTENSION TO 30 KB

Single cylinders operating under elastic conditions are limited to about 17 kb even after autofrettage, although up to 23 kb may be contained if a certain degree of permanent deformation is permissible. At these comparatively high pressures tungsten carbide is a preferred material since its elastic deformation is about three times less than that of steel but it is too brittle to use unless supported by steel rings. The use of multi-binding rings, either shrunk-on by heating or pressed-on against an interference fit, is considered in Chapter 6 in connection with pressure generation to 100 kb.

An alternative method of support is to use a cylinder of conically shaped outer surface and press it into a similar conical aperture in

48

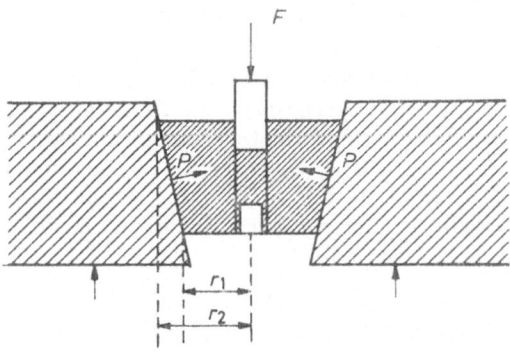

Figure 3.19. Supported cylinder (after Bridgman)

a large supporting block (*Figure 3.19*). The amount of external support to the inner cylinder is varied by changing the force driving the two together. From the geometry of *Figure 3.19* the supporting

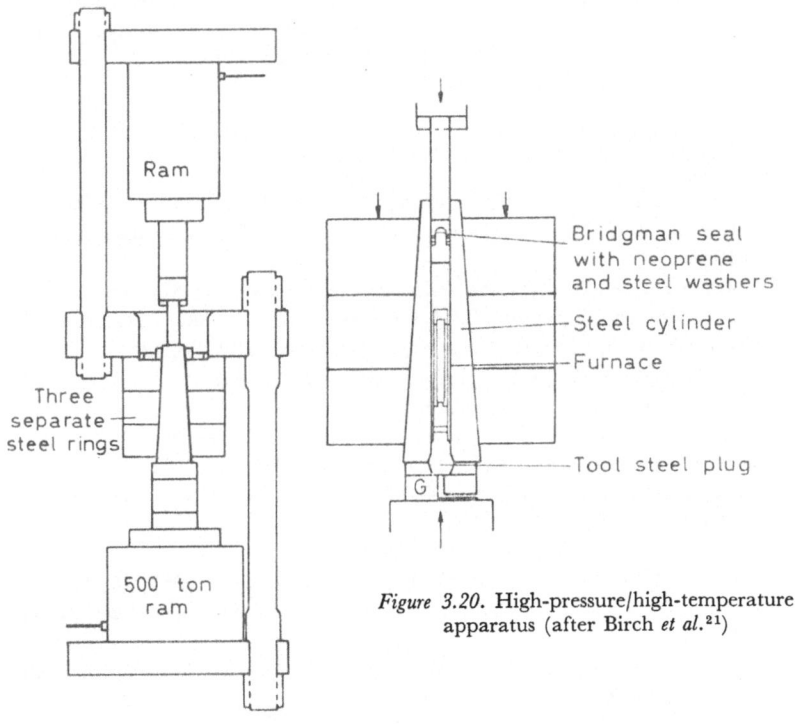

Figure 3.20. High-pressure/high-temperature apparatus (after Birch et al.[21])

pressure P is equal to $F/\pi(r_1^2 - r_2^2)$ where F is the total thrust. The latter can be the load applied to the piston in the inner cylinder which means that there must be a definite relation with the radii for the desired support. Alternatively the thrust on the cylinder can be applied completely independently of the piston load. The limiting external pressure is given by the strength of the outer cylinder.

This method, originally due to Bridgman[20], has been used by a number of people to generate pressures up to 30 kb and beyond and an example is given below.

Apparatus for Pressures to 27 kb and Temperatures of 1,400°C

The apparatus shown in *Figure 3.20* is due to Birch *et al.*[21] and has been used for geochemical synthesis. The supporting block consists of three rings of steel (R.C. 50). The tapered cylinder with 16° included angle and 1·8 cm bore is made from the same material. Load is applied to the piston and tapered cylinder by two independent intensifiers as shown. The piston seal is a Bridgman mushroom head packing and electrode seals of the type shown in *Figure 3.12* enter through a high-speed tool steel end plug. Either pentane or nitrogen is used as the pressure medium.

REFERENCES

[1] Bridgman, P. W. *The Physics of High Pressure*. Bell, London, 1958.

[2] Hamann, S. D. *Physico-chemical Effects of Pressure*. Butterworths, London, 1957.

[3] Comings, E. W. *High Pressure Technology*. McGraw-Hill, New York, 1956.

[4] Lamé, G. and Clapeyron, B. P. E. *Mem. prés. Acad. Sci.*, Paris, 1833, **4**.

[5] Crosland, B. and Bones, J. A. *Engineering*, 1955, **179**, 80, 114.

[6] Manning, W. R. D. *Ind. Eng. Chem.* 1957, **49**, No. 12.

[7] Patterson, M. S. *J. sci. Instrum.* 1962, **39**, 123.

[8] Whalley, E. and Lavergne, A. *J. sci. Instrum.* 1959, **36**, 46.

[9] Bridgman, P. W. *Proc. Amer. Acad. Ants. Sci.* 1937, **72**, 171.

[10] Reeves, L. R., Scott, G. J. and Babb, S. E. *J. chem. Phys.* 1964, **40** (12), 3662.

[11] Dadson, R. S. *Nature, Lond.*, 1955, **176**, 188.

[12] Dadson, R. S. and Greig, R. G. P. *Brit. J. appl. Phys.* 1965, **16**, 1711.

[13] Darling, H. E. and Newhall, D. M. *Trans. Amer. Soc. mech. Engrs*, 1953, **79**, 311.

[14] Bonen, M. D., Babb, S. E. and Scott, G. J. *Rev. sci. Instrum.* 1965, **36**, No. 10. 1456.

REFERENCES

[15] Kennedy, G. C. and La Mori, P. N. *Progress in Very High Pressure Research*, p. 304. Eds. F. P. Bundy, W. R. Hibbard and H. M. Strong. Wiley, New York, 1961.

[16] Gugan, D. *J. sci. Instrum.* 1956, **33**, 160.

[17] Blosser, L. G. and Young, H. S. *Rev. sci. Instrum.* 1962, **33**, 1007.

[18] Cornish, R. and Ruoff, A. *Rev. sci. Instrum.* 1961, **32**, 639.

[19] Fishman, E. and Drickamer, H. G. *Analyt. Chem.* 1956, **28**, 80.

[20] Bridgman, P. W. *Proc. Roy. Soc.* 1950, **A203**, 1.

[21] Birch, F., Robertson, E. C., and Clark, S. P. *Ind. Eng. Chem.* 1957, **49** (12), 1965.

[22] Heydemann, P. *J. appl. Phys.* 1967, **38**, 2640.

4

OPPOSED ANVIL APPARATUS

The serious consideration of extending pressure ranges up to and beyond 100 kb was first taken up by P. W. Bridgman at Harvard. He built a number of devices based initially on piston and cylinder configuration but with a number of modifications to improve the strength. The highest pressure obtained by him in this type was 100 kb in a two-stage assembly consisting of an inner piston and cylinder subjected to 30 kb external hydrostatic pressure inside a larger vessel[1] (see page 123).

A further account of multi-stage devices is given in Chapter 6.

To return to the subject of relatively simple high-pressure apparatus, a major problem is in increasing the usable limit of the piston. Obviously it cannot be supported over its whole length because of the finite stroke when under load. A material with one of the highest compressive strength is tungsten carbide and even pistons constructed from this have a very high failure rate above 50 kb. It was from a desire to improve this state of affairs that Bridgman developed an extremely simple apparatus (at least in construction) which can be used to 200 kb and above[1]. The principle is that of pressure generated between two opposed pistons or anvils by a uniaxial force without a containing cylinder. If the high pressure is generated on a small area of a much larger body it may be shown that two or three times the compressive yield stress can be supported because of its attachment to the surrounding material. Bridgman called this the principle of massive support. In *Figure 4.1* for example, the stresses set up between the small faces of the cones are fanned out to the main bulk behind them and hence the ultimate strength is increased. The gain in strength depends critically on the angle of the cone and it will be shown later that only semi-angles greater than 75° are really effective (see page 88). The theory of this principle may be found in a number of books but only the practical implications will be considered here.

Apparatus using two opposed pistons in the above manner is usually known as 'Bridgman anvils' and will be referred to as such in this book. It is essentially a 'two-dimensional apparatus' since under load the sample cell is pressed out so that thickness-to-area

52

ratio is of the order of 1:50. This is its main disadvantage since only extremely thin and small samples can be considered and for this reason a number of multi-anvil versions have evolved over the last ten years using three-dimensional pressure cells with volumes of several cm. The most commonly used are tetrahedral and cubic configurations which are described in greater detail in the next chapter. The usefulness of Bridgman anvils for simple measurements of resistivity changes under pressure is immediately apparent and in succeeding sections of this chapter a description is given incorporating some later modifications by other experimenters together with more sophisticated adaptations used to measure optical, x-ray diffraction and magnetic properties.

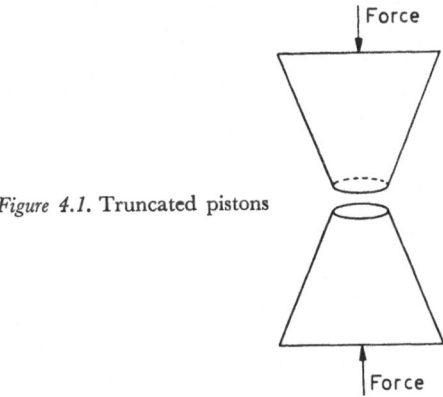

Figure 4.1. Truncated pistons

A number of theoretical solutions of the pressure distribution inside a material between Bridgman anvils have been made (see for example Jackson and Waxman[2]). These depend on whether the sample cell is plastic or elastic, contained by an outer ring or whether the anvils yield, etc. The calculations are not very useful for real problems except possibily in the case of samples deforming plastically between diamond anvils which may be assumed to be rigid.

Bridgman Anvils for Pressures up to 200 kb

Originally Bridgman used single pieces of tungsten carbide but most later designs incorporate supporting steel rings. It is important that the pieces of tungsten carbide should be as small as possible since the probability of finding flaws in commercially supplied pieces is obviously much greater the larger the piece used. The design of a complete apparatus is shown in *Figure 4.2* where both low-pressure and high-pressure versions are given. In the case of the high-pressure version the anvils consist of inserts of tungsten carbide (6 per cent cobalt binder) 1·27 cm in diameter and 2 cm long with 0·375 cm diameter flats ground on one of the plane faces at a 5° taper angle.

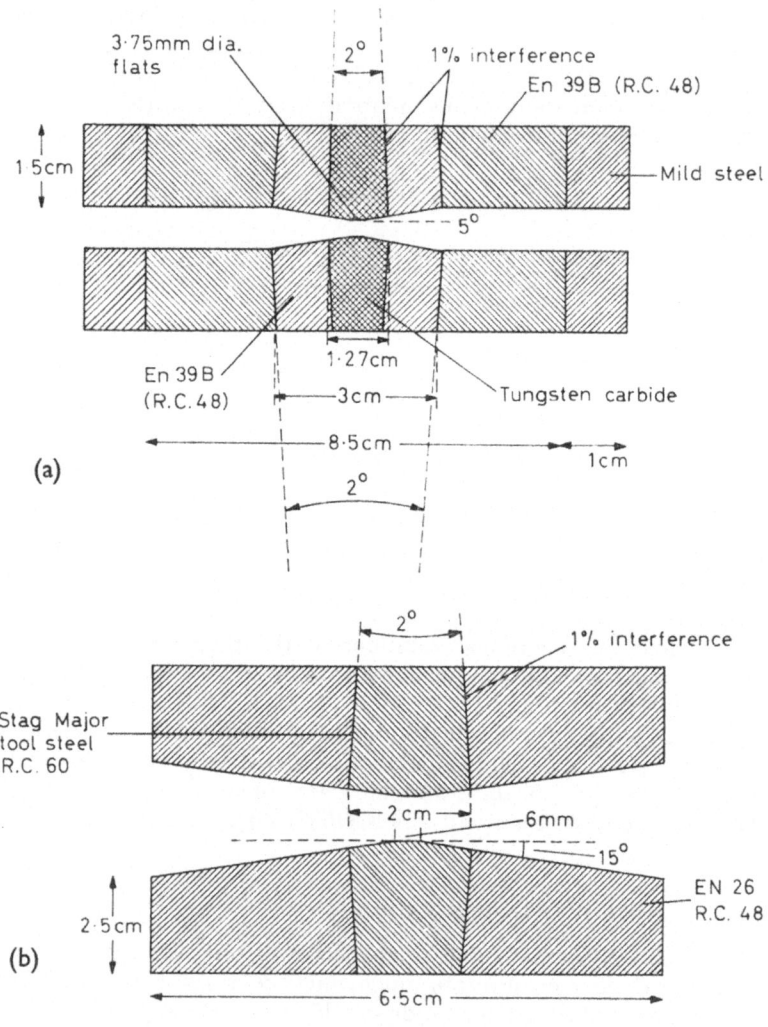

Figure 4.2. Bridgman anvils

54

With opposed anvil devices it is essential that the two high-pressure bearing flats are pressed together as parallel as possible otherwise the failure rate is very high. *Figure 4.3* shows an axial loading shackle

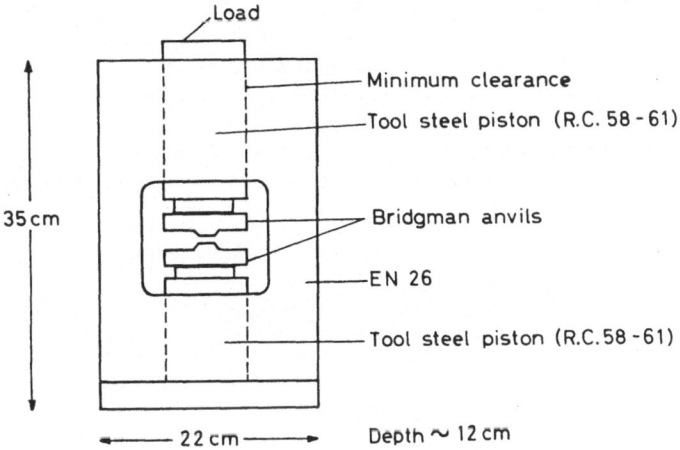

Figure 4.3. Axial loading shackle for Bridgman anvils

which may be used to ensure that this condition is fulfilled. The bore of the shackle is machined so that there is minimal clearance between the pistons and the yoke. The whole unit is then placed in a conventional press and loaded appropriately. Electrical insulation between the upper and lower anvils is maintained by inserting thin mica, or mylar discs. In this form the device can be surrounded by a furnace and heated to 500°C. (The strength of tool steel and tungsten carbide falls with temperature and limits the pressures to 50 and 70 kb respectively at 500°C.) The anvils may be cooled down to liquid nitrogen temperatures by installing cooling rings through which liquid nitrogen is pumped. Before tackling the problem of temperature well above and well below ambient the reader is advised to consult Chapter 2 and Appendix A, where the relevant properties of steels are listed.

In *Figure 4.2a* the anvils are strengthened by a double steel ring supporting system. (A theoretical analysis of the mechanism of supporting rings has been given by Chrisstiansen *et al.*[3] and is discussed further in Chapter 6.) Both rings are of EN 39 B steel hardened and tempered to Rockwell C 48 and are in 1 per cent interference. This means that before assembly the inner diameter of an outer ring is 1 per cent smaller than the outer diameter of the ring it is being fitted to. The rings are pushed together in a wedge profile 1° included semi-angle and molybdenum disulphide is used as a lubricant. The assembly is from the outside inwards and a surrounding mild steel safety ring is incorporated. At low pressures the tungsten carbide can be replaced by a Stag Major tool steel insert (R.C. 58–60) with a single support ring. As shown in the *Figure 4.2b* the flats are 0·6 cm diameter on a 15° taper.

The design of a sample cell to be placed between the anvils depends on the type of material and the particular property to be studied. For the moment an electrical resistance cell first used by Bridgman will be considered, and is shown in *Figure 4.4a*, where

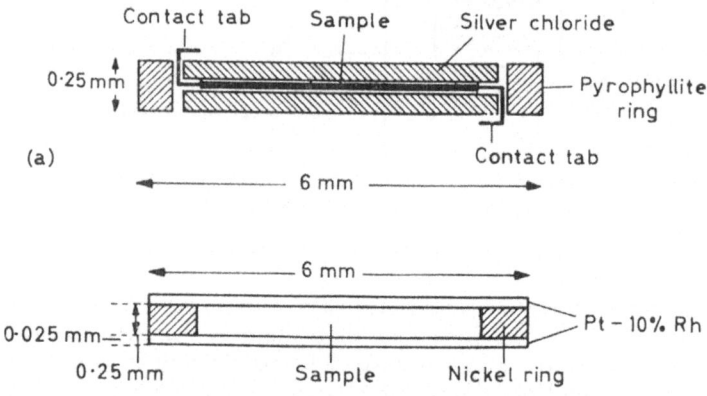

(b)   *Figure 4.4*. Sample cells for Bridgman anvil apparatus

the sizes are appropriate to the anvils shown in *Figure 4.2b*. Solid silver chloride has a low resistance to shear and when used as a pressure transmitting medium performs approximately hydrostatically. Under load it is contained between the anvils by a pyrophyllite or pipestone ring. Both these latter substances are normally soft solids, are easily machined to a required form and shear easily at low stress. Due to their microcrystalline structure considerable friction is generated both internally and at the interface with the anvils which limits the extrusion as load is increased thus sealing in the remainder of the resistance cell. The frictional properties are enhanced by coating the anvils, at the edges only, with jewellers' rouge or diamond dust which are applied conveniently from suspensions in acetone.

Pyrophyllite and pipestone are two naturally occurring minerals and are basically aluminium silicates. The former is the most commonly used since it has the better properties in terms of uniformity of grain structure and may be obtained from a number of sources (see Appendix B). Pyrophyllite is a very good electrical and thermal insulator but is porous even under load and cannot be used for containing liquids.

The sample under test in the form of a thin disc is sandwiched between two pieces of silver chloride with copper tabs as electrical connections. Contact to the anvils is either direct or via copper or gold discs freshly amalgamated. In general it is not possible to take leads out through the side of the pressure cell since this causes extrusion of the silver chloride and sample. Many materials are attacked chemically by silver chloride and it is desirable to coat those parts of the sample exposed to it by thin epoxy resin varnish or a similar substance. Silver chloride has a volume discontinuity at 90 kb but since this is only about 2 or 3 per cent there is usually no noticeable effect, at least when experimenting with fairly compressible samples.

Other pressure assemblies have been used, for example Myers et al.[4] have studied the effect of pressure and temperature on minerals contained by nickel annuli between Bridgman anvils (*Figure 4.4b*) and the above description has been included to give only a general guidance.

Pressure Calibration

The question of pressure calibration in Bridgman anvil apparatus is quite complicated and a major drawback in accurate work. Bridgman[5] first noticed that above the bismuth I–II point at 25·4 kb transitions determined by electrical resistance measurements in opposed anvils occurred at much higher apparent pressures than those obtained from volumetric measurements in piston and cylinder apparatus where the pressures are known with a fair degree of certainty. For example, a transition in barium was at 80 kb in the first case and 59 kb in the second. In this context the pressure in opposed anvils is defined as simply total force over area. Since only a negligible degree of plastic deformation of the anvils occurs below 100 kb Bridgman inferred that the true pressure was very close to this mean value which left the above anomalies unexplained. It was observed that at fairly high pressures a lens-shaped deformation occurred but since this probably gives rise to smaller areas being in contact it would make the real pressure considerably greater than the mean one which is contradictory to the observation above. The equivalence of the resistance and volume transitions in bismuth, barium, thallium etc., has been demonstrated by a number of

57

people[6,7] which indicates that the discrepancy in the opposed anvils arises from behaviour peculiar to this configuration.

It should be emphasized at the outset of any discussion of this calibration that the results of one particular set of experiments can in no way be extended even in very general terms to others. The experimenter must calibrate always in detail his own device. For example, it has been shown that pressure gradients across the anvils may be zero, positive or negative depending on the nature and thickness of sample and pressure medium, anvil size, temperature and any other variable parameter. In the method of Montgomery et al.[8] bismuth and manganin wires are used as pressure indicators. The samples are placed in a number of geometrically different positions between Bridgman anvils and the appropriate real and apparent pressures determined from a previous knowledge of the pressure dependence of their electrical resistance. The two phase transitions in bismuth which occur at 25·4 and 89 kb on the usually accepted pressure scale (see Chapter 1) are used together with a manganin resistance/pressure relation extrapolated from hydrostatic experiments. A calibration is obtained which is accurate to 1 or 2 per cent at 125 kb. In order to determine the pressure profile across the anvils, bismuth wires are placed vertically at different distances from the centre in a silver chloride cell of the type shown in *Figure 4.4a*. The results for 1·27 cm diameter anvils and 0·25 mm thick pressure cells indicate that pressure is lowest at the centre and highest near the edges with an approximately linear gradient. This varies considerably with the diameter of the flats and thickness of the pressure cell. It seems probable that at the outermost edges of the anvils the pressure is only a few kilobars since this is the limiting shear stress for pyrophyllite and moving inwards towards the centre the pressure rises rapidly for a short distance and subsequently falls towards the centre due to the inability of silver chloride to follow the stress pattern of the anvils arising from the lens-shape deformation as observed by Bridgman. Hence samples placed in the central regions will be subjected to pressures less than the mean value, which explains to some extent the anomaly observed by Bridgman. In addition to these measurements the resistance of manganin wire arcs of different radii placed in a horizontal position in the cell is determined and an extrapolated pressure scale obtained.

It is found that at high temperatures the gradient across the anvils is less, due presumably to the greater ease with which silver chloride takes up the stress distribution. Other experimenters have studied bismuth and thallium transitions in much thicker cells and have found a greater pressure at the centre than at the edges which emphasizes the need for calibrations *in situ*.

Another method of calibration which is mainly useful for a combination of high temperatures and rather lower pressures than those considered above is used by Myers *et al.*[4] In this a known phase diagram is used, for example the synthesis of coesite from quartz in the presence of water or the analogous crystabolite to quartz transition in boron phosphate or arsenate. The coesite/quartz phase diagram used by Myers *et al.* is shown in *Figure 4.5.* (Takahashi[9]

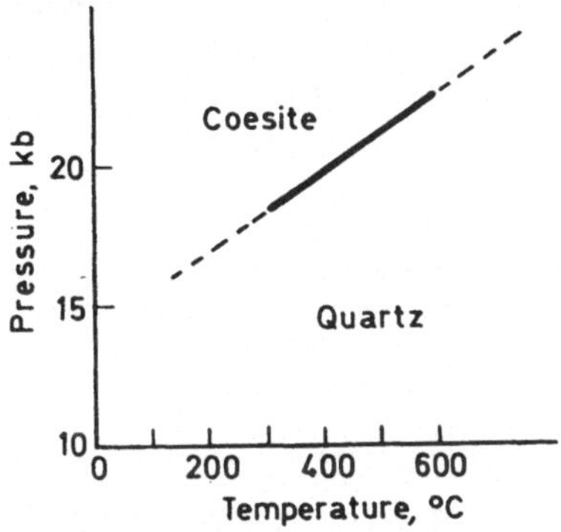

Figure 4.5. Pressure/temperature relation for the quartz/coesite synthesis in the presence of water

has shown that this phase diagram may not be reliable as it disagrees with other experiments in piston and cylinder and multi-anvil apparatus.) High temperatures may be generated in Bridgman anvils most conveniently with a surrounding external furnace since the high thermal capacity and conductivity of usual anvil materials make internal heating difficult. The sample cells consist of polycrystalline powder contained in a nickel annulus and sandwiched between two platinum rhodium sheets (*Figure 4.4b*). The temperature is monitored by a thermocouple placed between the anvils. After the application of a given load and temperature the sample is quenched to normal conditions and an x-ray diffraction method used to determine the synthesis. Myers *et al.*[4] varied the ratio of the ring diameter to the sample diameter and the results indicated that in this particular assembly the pressure is uniform across the anvils at high temperatures. This method provides a calibration between 20 and 60 kb and 0 and 600°C but the unreliability of phase diagrams is a major disadvantage.

In a number of laboratories[4,10,11] the densification of silica glass has been used as a pressure monitor. Independent determinations

59

have been made of the density and refractive index of glass as functions of pressure at different temperatures. The samples are placed in noble metal capsules and are quenched to room conditions after the required pressure and temperature cycle. Unfortunately considerable doubt must be placed on this method of calibration because of the dependence on the degree of shear and time taken for equilibrium to be reached.

A third method of calibration has been suggested by Bradley *et al.*[12] In this the sample constitutes an electrolytic cell and measurements are made of the e.m.f. development as a function of temperature and pressure. This is easily related to the change of free energy of the system and therefore to pressure and volume. A suitable arrangement is shown in *Figure 4.6*, but the disadvantage is that only a mean

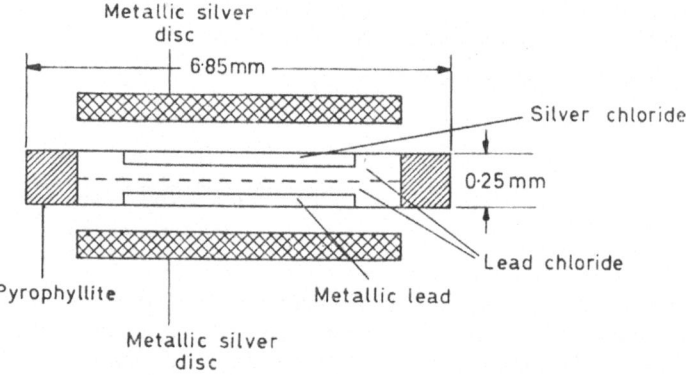

Figure 4.6. Pb/PbCl₂|AgCl/Ag electrolytic cell (after Bradley, Haygarth and Munro[12])

pressure can be measured and the presence of an electrolytic cell in a sample assembly would tend to disturb the pressure distribution.

A fourth method is to measure phase transitions or crystal lattice parameters by x-ray diffraction measurements. This technique will be described in greater detail in following sections of this chapter.

In summarizing the generation of high pressure in simple Bridgman anvils it could be said that for conventional resistance experiments, particularly for determining phase transitions, they are an extremely useful device up to 200 kb or rather lower pressures at elevated temperatures. Calibrations may be made fairly easily to 100 kb with an accuracy of a few per cent, always bearing in mind that the absolute pressure scale is not well established in this range.

The main difficulty lies in assembling test cells with similar construction to the ones in which calibrations have been made or in placing a calibrant in a cell so that it does not appreciably affect the pressure distribution. For the highest pressures (above 200 kb) allowance has to be made for deformation of the anvil faces, for example Montgomery et al.[8] found that at 300 kb the true pressure can be 15 per cent down on the pressure calculated from the original area.

Bridgman Anvils Used for X-ray Diffraction up to 500 kb

X-ray diffraction experiments in conventional high-pressure apparatus, that is piston and cylinder, require basically a construction material which is transparent to x-rays (usually molybdenum radiation), which therefore limits the choice to those based on elements with low atomic number (usually less than 6) for example beryllium, carbon (diamond), and sapphire (aluminium oxide). Piston and cylinder cells of beryllium and diamond have been built and used very successfully (usually with steel pistons). Vereshchagin[13] has used beryllium cylinders with steel supporting rings up to 20 kb and Kasper[14] has employed a single diamond crystal with a small cylindrical hole ground in it. The disadvantages of these are that in one there is excessive plastic behaviour above 20 kb and in the second the smallness of the cell due to cost and availability leads to sealing difficulties and appreciable friction.

Bridgman anvils are an obvious choice for an x-ray diffraction experiment up to very high pressures since only small samples are needed usually and careful focusing of the x-ray beam can obviate the effects of the anvils. The sample cells shown in *Figure 4.4* are obviously unsuitable for x-ray diffraction because of the considerable number of lines from pyrophyllite and silver chloride. Hence the main requirement is a retaining medium which is only weakly diffracting.

The Jamieson and Lawson[15] type of apparatus shown in *Figure 4.7* follows the usual design of Bridgman anvils and has in addition a collimator for molybdenum radiation and a camera radius 9 cm covering an angle of 120°. Load is applied by an hydraulic ram on to 3 mm diameter flats on the tungsten carbide anvils, the multiplication factor being about 120. The sample is contained in an amorphous boron wafer about 0·75 mm thick which has an x-ray diffraction pattern resembling a smudged-out ring. The wafer is formed on the lower anvil by hand compacting using a thin glue solution as a binder and the sample is placed in a 0·375 mm hole drilled centrally. Under compression the wafer reduces to about

0·25 mm. It is found that there is little trouble with reflection from the anvils and good diffraction patterns are obtained with normal exposure times. The pressure distribution across the anvils is such that the pressure at the centre is two or three times the mean value. Although a theoretical equation can be used to determine the

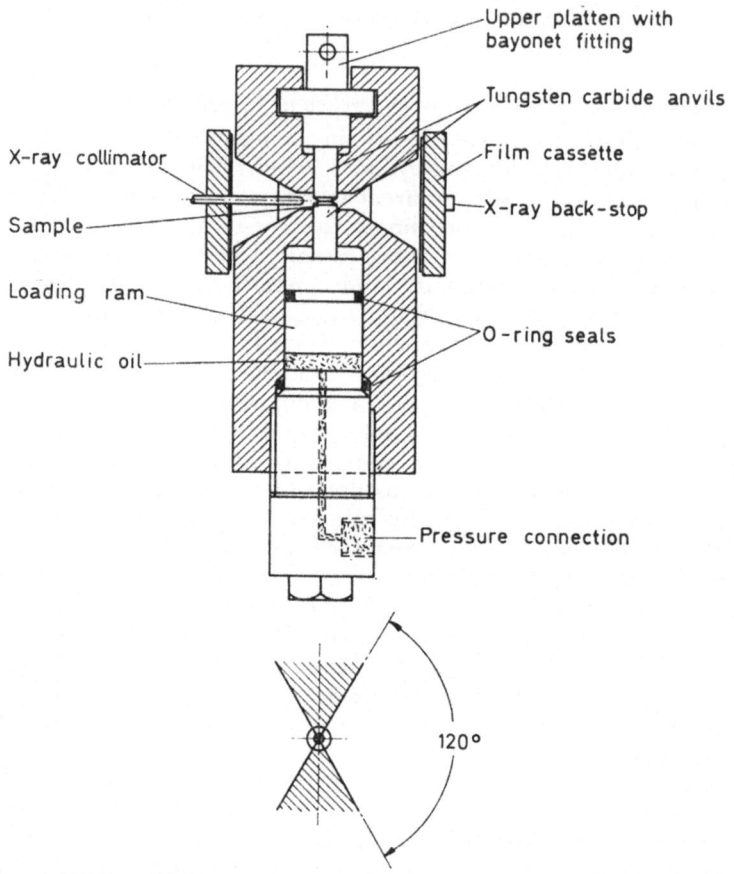

Figure 4.7. X-ray high pressure apparatus for 9 cm camera (after Jamieson and Lawson[15])

calibration approximately, a more straightforward procedure is to monitor phase changes and lattice parameters in substances such as silver chloride or sodium chloride which are known from other experiments. The highest pressures reached are about 150 kb (actual not mean).

Recently Jamison[16] has described modifications to enable x-ray diffraction experiments to be carried out at temperatures between −100°C and +350°C. The method is by directly heating or cooling the anvils. Temperatures are measured as closely as possible to the sample by using anvils of composite structure to avoid drilling holes in tungsten carbide. Heating may be carried out by another method in which pyrolitic graphite is mixed with the sample and boron so that a few ohms resistance is obtained across the anvils. The graphite has very little detrimental effect on the diffraction patterns and temperatures of 350°C can be obtained.

National Physical Laboratory High-Pressure X-ray Diffraction Apparatus

Miniature diamond anvils offer a number of advantages, namely the extremely hard nature of the diamond lattice and the fact that with single crystals any scattered radiation is in the easily recognizable form of Laue spots which do not confuse the Debye–Scherrer powder pattern of the sample. Although diamond anvils have been used mostly for polycrystalline experiments it is possible to make single crystal measurements as will be described later.

The high-pressure x-ray device developed at the National Physical Laboratory by Owen and Smith[17] is designed to fit on to either a Hilger and Watts x-ray camera 5·73 cm film diameter or a Unicam camera 11·46 cm diameter. The respective angles of rotation available are 120° and 300°. The detailed assembly of the apparatus is shown in *Figure 4.8*. In operation the sample is squeezed between a pair of miniature diamond anvils and a finely collimated x-ray beam incident at right angles to the axis of the anvils, that is at right angles to the pressure axis, is used to obtain powder diffraction patterns. It is found that self-friction in the sample is usually sufficient to retain a small amount between the anvils under load without a constraining gasket.

The anvils are derived from commercially available Rockwell hardness indentors which consist of a diamond firmly embedded in the end of a steel piston. These are adapted by grinding down the piston to 6 mm outer diameter and polishing a flat between 0·125 and 0·75 mm diameter on the end of the diamond tip. The included cone angle is 120°. The actual sizes depend on the requirements of the experiment. Two of these indentors used as Bridgman anvils can withstand loads up to 250 kg before failing which is equivalent to pressures in the 200 kb region on 0·375 mm flats. The diamonds are small pieces silver soldered into conical recesses in the steel piston and the eventual failure at high loads can be attributed to uneven loading on the rear surface causing them to tilt and fracture. Owen and Martin[18] have modified the indentors further to overcome this failing at comparatively low pressures. The anvils are made from tool steel and octahedral and dodecahedral stones which are of the highest grade obtainable, that is with a minimal number of flaws. The stones are approximately 1/3 of a carat (1/15 of a gram), so that an adequate flat bearing surface

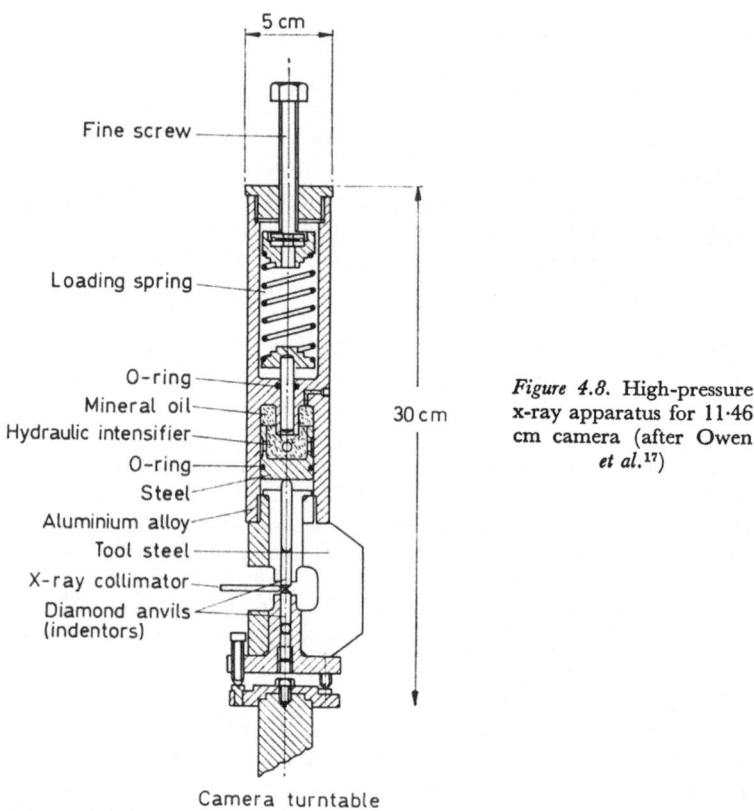

Figure 4.8. High-pressure x-ray apparatus for 11·46 cm camera (after Owen *et al.*[17])

may be polished on the rear face. They are orientated so that the pressure axis is parallel to the [100] direction in an octahedral crystal, the [110] direction in a two-point dodecahedral crystal and the [111] direction in a three-point dodecahedral crystal. A number of different seatings for the diamond have been tested and two are illustrated in *Figure 4.9.* In (*a*) the diamond is seated directly onto a flat surface turned in the steel piston, or a piece of tungsten carbide

64

may be interposed between the diamond and the flat surface and in (*b*) the diamond is mounted on a spherical ball seating. In method (*a*) a hot-setting technique using Johnson Matthey easy-flow silver solder is employed to retain the stone and in method (*b*) the seating is preformed by loading the half-ball to 450 kgf prior to the final assembly and polishing the diamond which is then cold set.

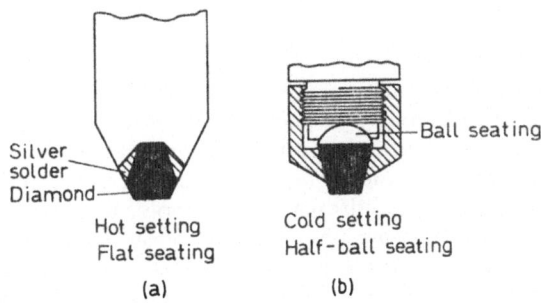

Figure 4.9. Seatings for diamond anvils (after Owen and Martin[18])

The anvils are used in pairs with the lower one about 0·125 mm greater diameter so that there is complete overlapping. They are accurately located in a tool steel frame (R.C. 55) as shown and load is applied to the upper one with a piston made from a similar material. The clearance in the cylindrical hole of the frame is less than 0·0075 mm. The apparatus can be loaded directly to 100 kgf by a helical compression spring. Alternatively the sliding piston is arranged to be an integral part of an oil intensifier through which the spring loading is magnified approximately twenty times. The spring holder and intensifier are constructed of aluminium alloy in order to keep the weight to a minimum and can take springs of different size and compressive strength. The applied loads are determined by noting the compression of the previously calibrated spring using a finely threaded screw. Provision is made for a small ball race between the springs and the nut used to apply the load in order to reduce rotational friction. Loads greater than 500 kgf have been applied to diamond flats 0·375 mm diameter giving mean pressures up to 500 kb. The limit is reached when the diamonds fracture and the most important factor which decides this point has been found to be the method of mounting rather than the orientation of the crystal. The most successful combination is the half-ball seating with either a two- or three-point stone since it appears that a slight alignment of the anvils is allowed in this seating while still maintaining load. The use of tungsten carbide inserts increases the pressure limit with octahedral stones by about 30 per cent.

It should be remembered that pressures up to 300 kb are easily reached without an excessive failure rate with just slight modifications of the commercially available hardness indentors.

Provision for winding heaters on the above device can be made and temperatures up to 300°C generated.

Operation

The high-pressure unit is mounted on the x-ray camera on a small turntable and stands freely on a cone and vee groove mounting. After the latter is accurately positioned with regard to the axis of rotation of the camera, the high-pressure device is removed and the lower diamond anvil withdrawn. The sample, usually in the form of a thin wafer of precompressed powder about 0·125 mm thick, is attached with glue to the lower anvil. This is then replaced and pressure applied. Surplus material which extrudes from between the anvils is removed in order to minimize the amount of sample giving rise to the diffraction of the x-rays. If pieces of material from single crystals rather than powders are used the resulting diffraction pattern is not as satisfactory because of preferred orientation. The pressures to which the sample is subjected are not known very accurately and in the first place are quoted mainly as force over area, the latter being the unstressed diamond flat area (see below).

Calibration

The pressures between miniature diamond anvils are not known very accurately because of the high gradients created in the self-friction retaining mechanism. There is some evidence that the pressure at the centre is three or four times the force over area value, for example the crystallographic change in indium antimonide which occurs at about 25 kb has been found at 8 kb mean pressure in this apparatus. The most influential factors in the pressure distribution are the type, thickness, shear strength and compressibility of the sample material, and the surface finish and parallelism of the diamond anvils.

To some degree it is more satisfactory to consider empirical methods for a quantitative assessment of the pressure gradient between the anvils rather than mathematical analyses. Lippincot and Duecker[19] have measured the optical absorption of nickel dimethylglyoxime diluted with various alkali halides between opposed anvils with areas between 0·01 and 0·001 sq cm. The distribution across the anvil surfaces of the absorption band at 19,000 cm^{-1} is determined with a spectrophotometer which can be focused on areas as small as 10^{-7} cm^2. The absorption band has a frequency shift of plus 80 cm^{-1} per bar approximately and the resulting distribution can be represented by a parabolic curve of the type

$$P_{(r)} = P_m \left[1 - \left(\frac{r_0}{r_0} \right)^2 \right]$$ P_m is the centre pressure, r_0 is the anvil radius

although the experimental results shown in *Figure 4.10* might indicate linear gradients.

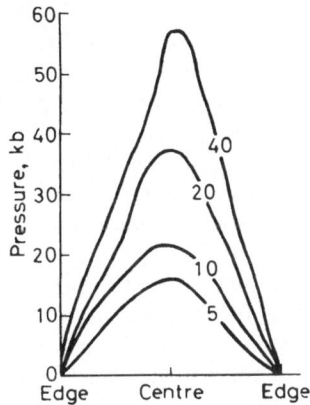

Figure 4.10. Pressure gradient for nickel dimethylglyoxime in KBr (1:2) at specified applied mean pressures (after Lippincott and Duecker[19])

There is a certain degree of generality in the approach of Lippincott and Duecker but the individual characteristics of different samples cannot be accounted for. If the above equation is integrated with respect to r and the answer put equal to the mean applied pressure it may be shown that the centre pressure is approximately a factor of two greater than the mean at low pressures, this factor decreasing with increasing pressure. To some extent this behaviour is general.

The N.P.L. x-ray apparatus has the considerable advantages of low cost, ease in operation and a high pressure limit of several hundred kilobars. This is counter-balanced to some degree by the presence of large pressure gradients giving sometimes two or more phases present simultaneously.

Lattice spacings obtained with molybdenum radiation are accurate to ± 0.03 Å at 300 kb. Copper radiation would provide an increased accuracy but generally speaking exposure or recording times are inconveniently long.

Experimental Results

The structure of new phases formed under pressure is in many cases more important than knowing accurately the transition pressure and the apparatus may be used to great advantage for this type of investigation because of its wide range.

The semiconducting III–V compounds based mainly in the third, fourth and fifth periods of the periodical table have zinc blende structures under normal temperature and pressure conditions. Electrical and optical absorption experiments have shown that they transform to new phases at pressures ranging from tens to hundreds of kilobars[25,29]. The β tin analogue of these compounds is interesting because of its position in the table. Martin and Smith[20] have examined tin and indium antimonide to 300 kb in the above apparatus and have shown that the latter transforms to a white tin structure (four nearest and two next nearest neighbours) at 25 kb which persists up to 250 kb where there is a further transformation to a caesium chloride structure. White tin transforms to a body-centred tetragonal structure at approximately 115 kb but appears to have no further phase changes up to 300 kb. In another series of experiments Martin and Smith have shown that in the isoelectronic series β Sn, InSb, CdTe and AgI the first two to transform to a white tin structure, CdTe goes to a sodium chloride structure and then to a white tin structure and AgI goes only to sodium chloride thus showing a gradual change of behaviour with the degree of ionic bonding.

Piermarini and Weir X-ray High-Pressure Apparatus

One of the major disadvantages of the two x-ray devices in *Figures 4.7* and *4.8* is that the x-ray beam passes through the sample laterally and hence scans the complete pressure gradient. One way of avoiding this, to a large degree, is to pass the beam through the diamonds along the pressure axis. If a finely focused beam is used only a small fraction of the anvil area is operative. The version shown in *Figure 4.11* is based on an original design of Piermarini and Weir[21] but with some modifications. One anvil is of the type used above, that is a converted hardness indentor and in this case the diameter of the steel shank is 0·8 cm and is drilled out to 3 mm tapering to 0·25 mm at the back face of the diamond. The second anvil is simply a small diamond (about 2 mm × 2 mm) with two parallel polished faces which is rung on to the optically polished surface of a steel supporting block. The beam enters the indentor anvil through a lead glass collimating tube and after traversing the sample between the diamonds passes through a 60° sector cut in the supporting end plate as shown. The available 2θ angle is 60°. Force is applied by a small hydraulic system as shown. The main parts are made from EN 26 steel hardened and tempered to R.C. 45 and with the sizes used oil pressure up to 40 b is necessary to generate about 60 kb on a diamond flat of 0·5 mm diameter. This is a practical limit because of the comparative weakness of the unsupported second diamond anvil.

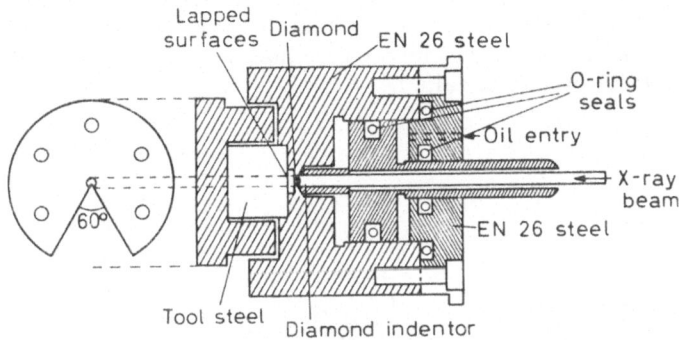

Figure 4.11. Diamond high pressure x-ray cell (after Piermarini and Weir[21])

Weir and Van Valkenberg Optical High-Pressure Apparatus

The transparency of diamond to radiation from ultra-violet to far infra-red frequencies with only a few absorption bands (page 111) makes it an obvious choice for high-pressure optical experiments using a similar configuration to that shown in the x-ray cell in *Figure 4.11,* that is where the radiation is transmitted along the axis of the applied load. In the device shown in *Figure 4.12* two gem quality type II diamonds each about 1/5 of a carat are used as anvils. They are polished with two parallel faces and set in stainless steel pistons which are constructed in two parts so that adjusting screws can be used to make the anvil flats accurately parallel. The observation of interference fringes with for example mercury green light is used for this purpose. A hole is drilled in the pistons and is conical up to 1·5 mm from the back of the diamond. A hardened steel insert which is itself fixed into the main steel block holds the pistons and is connected to a lever system for generating pressure from a compressed spring. The mechanical parts are arranged so that there is no force on the diamond anvils other than at right

69

angles to the flat faces. The specimens are placed directly between the anvils as for x-ray diffraction and are either pure or are diluted with alkali halides if they have absorption bands which are too intense. Lithium fluoride is the best choice for a diluent with regard to resistance to extrusion but it has an absorption band at 14·5 microns and reacts sometimes with the sample. Provision for heating may be made by cutting channels in the main block and circulating fluids. The diamonds may be replaced by sapphires for pressures up to a few tens of kb but they are very much less use optically. This high pressure cell has the advantage that its operative dimension is only about 2·5 cm and it can be placed easily in normal spectrometers and microscopes. The pressure limit is approximately 200 kb (mean force over area value).

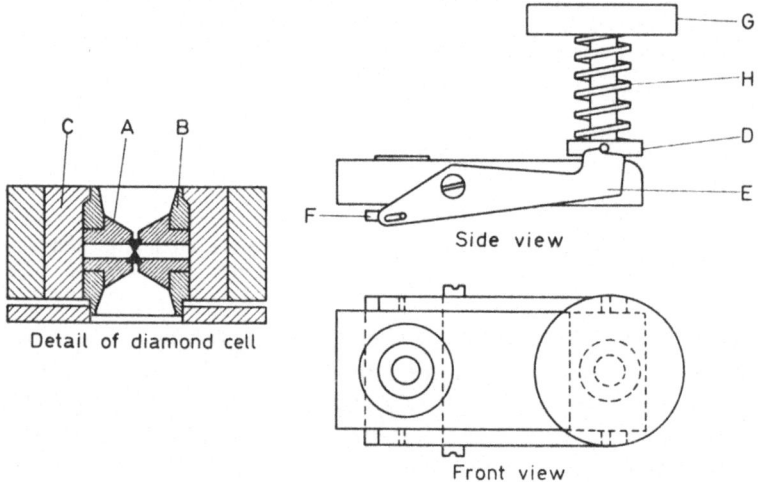

Figure 4.12. Diamond anvil high pressure optical cell. A and B, parts of piston; C, hardened steel insert; D. presser plate; E, lever; G, screw; H, calibrated spring (after van Valkenberg[24])

Calibration of Weir and Van Valkenberg Device

The pressure is usually determined by noting the compression of the spring when solid samples are used. For the reasons already given this is unlikely to be an accurate assessment. Connell[23] has examined Newton's ring fringes with mercury green light for different materials between the anvils. The pressure profiles are determined by the bunching of the fringes and it is found that this depends on the compressional yield stress of the sample. If it flows plastically with ease then the highest pressures are at the centre, conversely if flow is difficult then the highest pressures are to be found at the edges. Using a zinc sulphide single crystal the transition to a sodium chloride structure observed optically begins at the edges since it is difficult to deform. A powdered sample, however, transforms firstly at the centre.

The optical high-pressure apparatus has been used by Van Valkenberg[24] to study crystallographic forms under pressure by polarization microscopy. Diamond is isotropic and therefore non-birefringent except for flaws. The technique is to punch out small discs of incovar 0·125 to 0·25 mm thick which are placed between the diamond anvils and a small pressure applied. They are then removed and holes 0·3 mm diameter drilled in the centre of the outline of the diamonds. The holes are filled with a suitable liquid medium and the crystal of interest. Under load a seal is made at the line of contact between the diamond and the metal. Optical constants are determined in the usual way with polarization microscopy. Other metals than incovar may be suitable as long as they do not appreciably work harden.

DRICKAMER'S SUPPORTED ANVIL DEVICES

Under this heading are included a series of devices based on the Bridgman anvil design in which the tapered surfaces of the anvils are supported by external means. They form a cross between opposed anvil and piston and cylinder apparatus and have been used for optical, electrical resistance, x-ray diffraction, nuclear magnetic resonance and Mössbauer resonance experiments by Drickamer and his colleagues at Illinois. The basic design is an electrical resistance cell capable of pressures up to 500 kb and for the other experiments detailed modifications are made to sample assemblies and pressure media. A description has been given in considerable detail by Drickamer and others[25,26] and the information given here is based on these sources with the exception of a low-temperature optical version built by Sherman[31], although the construction materials are limited to those used generally throughout this book.

Electrical Resistance Cell

The high-pressure electrical resistance cell is shown in *Figure 4.13* and consists of a tungsten carbide insert supported by a steel ring (EN 26 R.C. 44–46) in 1 per cent interference fit. The anvils are also of tungsten carbide with steel supporting rings and are 3 cm long and 2·25 cm diameter with 0·25 mm diameter flats forming truncated cone ends with 18° taper. One piston is reduced in diameter by 0·100 mm so that an insulating mica sleeve can be placed around it. The profiles of the bottom piston and lower part of the tungsten carbide insert are as shown.

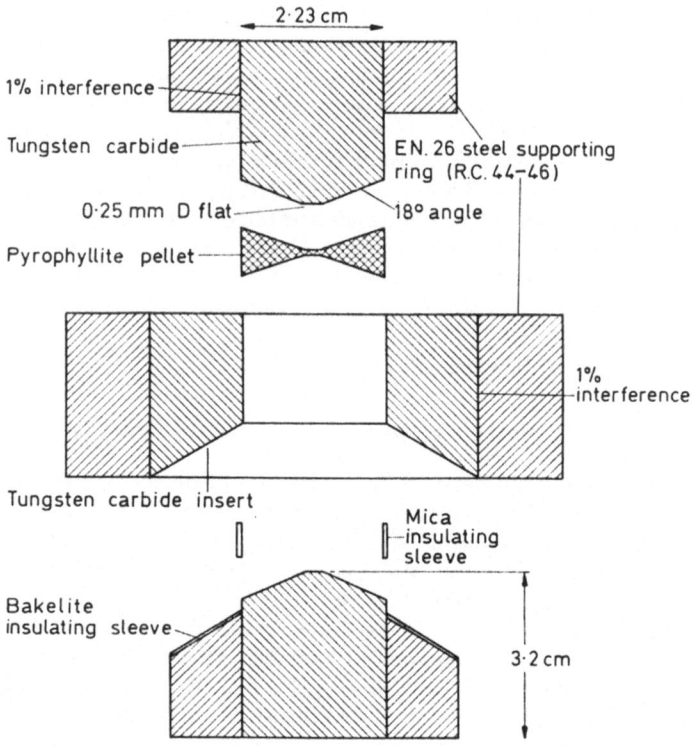

Figure 4.13. Supported anvil resistance apparatus (after Balchan and Drickamer[25])

Support for the anvils at the high pressure bearing faces is provided by a plug of pyrophyllite suitably shaped so that it is about 0·3 mm thick at the centre (all sizes given here relate to the apparatus size given in *Figure 4.13*). In operation the pyrophyllite pellet is compressed initially to 2 kb average pressure (average pressure refers to the total pressure supported by the 2·23 cm diameter anvil). The central flat area is removed with a flat-ended drill and the cell assembly shown in *Figure 4.14* placed in position. This is similar to that shown in *Figure 4.4a*. It is recommended that the total thickness of the inner components is about 0·01 mm less than the thickness of the ring of pyrophyllite. It is important for consistent pressure–load characteristics to control the centre thickness to the order of 0·003 mm. Measurements are taken from the point where the centre has formed into a single conglomerate which may be as high as 40 kb central pressure.

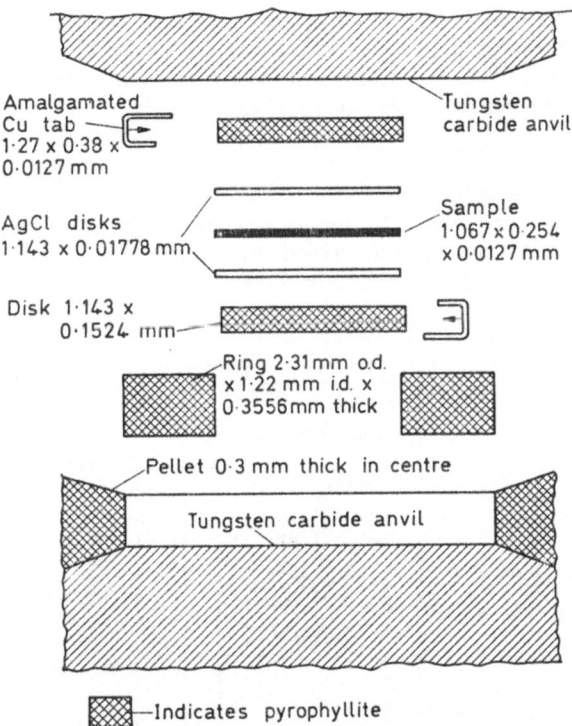

Figure 4.14. Centre assembly of high pressure resistance cell
(after Balchan and Drickamer[25])

The successful use of the apparatus to 500 kb is accomplished by work hardening the tungsten carbide anvils. This is carried out by compressing a pellet of pyrophyllite in the centre to an average pressure of up to 25 kb and regrinding to the original profile after the subsequent plastic deformation. Balchan and Drickamer[25] recommended 14 kb for work hardening tungsten carbide with 6 per cent cobalt binder and 25 kb in the case of 3 per cent cobalt binder.

73

In most cases both anvils are totally destroyed after reaching 200 or 300 kb central pressure but exceptionally some remain intact and with these pressures up to 500 kb can be obtained. It is found that other materials than pyrophyllite used for the support medium extrude too rapidly and anvils with tapers greater than 18° cause excessive extrusion and with less than 18° need very high loads for a given pressure[26]. Increasing the centre thickness beyond 0·30 mm in the above cell results in a very rapid increase in the load required for a given pressure but there appears to be no advantage in using thicknesses lower than this. Using the above techniques there is little or no evidence for appreciable pressure gradient across the high-pressure-bearing flats except at the edges[25].

Press

The press used by Drickamer for most of the work with the resistance and other cells to be described later is an intensifier similar in construction to one described in Chapter 3 (page 45). As shown in *Figure 4.15* the main body is about 15 cm o.d. and

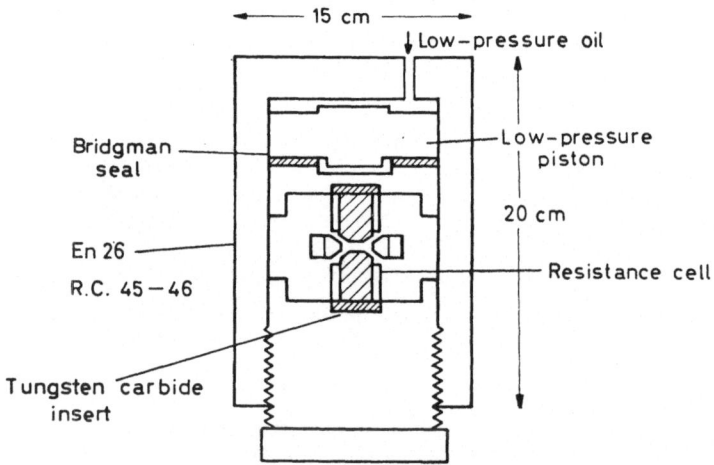

Figure 4.15. Intensifier for use with resistance and optical cells (after Fitch, Slykehouse and Drickamer[29])

12 cm i.d. and is about 20 cm high, machined from EN 26 steel hardened and tempered to R.C. 44–46. In the low-pressure piston side the seal is an unsupported area type and there is a connection to a low-pressure oil pump. Windows 5 × 6·25 cm are cut to coincide with exit and entrance ports of the optical cells (see below).

Calibration

Calibration of the electrical resistance cell has been studied in detail by Balchan and Drickamer[25] and has led to the suggestion of a number of new fixed points above 100 kb (see Chapter 1). The electrical resistance transitions in bismuth at 89 kb* and in barium at 59 kb* are determined, and the resistance of lead, platinum and indium measured as a function of the applied load on the anvils. (These metals are chosen because Bridgman's experiments[27] to 30 kb indicated slowly varying resistance/pressure relations.) Using Bridgman's data and the pressure volume data to several hundred kb from the shock wave work of Rice *et al.*[28], extrapolated functions of relative resistance versus pressure for lead, indium and platinum are calculated and hence a load versus central pressure relation for the cell is obtained. In the case of lead it is found that this is linear to the point where there is a transition at 161 kb and the consistency of this point using the different resistance relations justifies to some degree the extrapolations above the bismuth transition at 89 kb.

The scale can be extended to 500 kb using the data for indium and platinum. The load versus centre pressure relation is linear to 250 kb but above this there is a gradual curving over due to plastic deformation of the anvils. However it has been shown that for sets of similarly hardened anvils a normalized relation referring to this point can be obtained and hence used for calibrations in the 300 to 500 kb region. It is important that experiments and calibrations are carried out successively in a given set of anvils.

Using this calibration the resistance of barium, calcium, iron, rubidium and tin have been studied to 500 kb. The figures given in *Table 4.1* show that in all of these there are either transitions or sharp discontinuities in resistance in this range and hence suggest that they may be used as fixed points (see Chapter 1).

Table 4.1 (after Balchan and Drickamer[25])

Element	Pressure (kb)	Character of change in resistance
Bi	90	Sharp drop in resistance 250–300%
Fe	133	Sharp rise in resistance 366%
Ba	144	Sharp rise in resistance 42%
Pb	161	Sharp rise in resistance 23·2%
Rb	193	Sharp rise in resistance 147%
Ca	$\begin{cases} 425 \\ 375 \end{cases}$	Maximum in resistance Maximum in resistance

* See footnote on page 4.

75

It must be remembered that the calibration is based on a gross extrapolation beyond 89 kb and although the transition observed in iron at 131 kb is in good agreement with other results, that is from shock wave and belt apparatus, it should serve only as a guide rather than a standard for other experimenters.

High-Pressure Optical Cell for Use up to 140 kb in the Spectral Range 0·25 to 10 microns

Optical absorption experiments beyond the 30 kb hydrostatic region become increasingly difficult because it is necessary to keep the size of apertures down to a minimum so that the pressure container retains its strength. The first apparatus to successfully overcome this is an adaptation of the resistance cell which has been described above[29]. A series of holes with gradually reducing diameters filled with fused sodium chloride are used to hold pressures up to 140 kb between tapered anvils. Since the basic form is similar to that shown in *Figure 4.13*, only details of the windows and pressure medium will be considered at length.

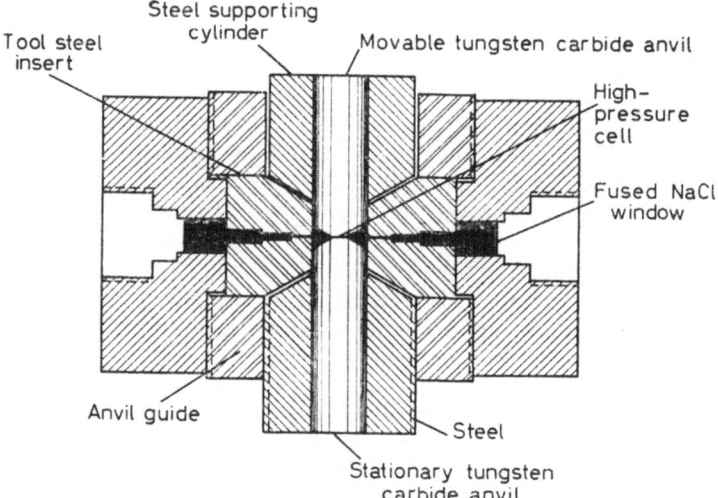

Figure 4.16. Supported anvil optical apparatus (after Fitch, Slykehouse and Drickamer[29])

[A simplified version capable of 50 kb has been used in a number of laboratories, the essential difference being that 3 mm pistons rather than tapered anvils are used[29].]

The high pressure version is shown in *Figure 4.16*. The central part of the cell is made from hardened tool steel (R.C. 58–60)

76

instead of tungsten carbide since holes are more easily fabricated in this material. EN 26 steel supporting rings are used as before. The profiles of the holes are as shown. The inner hole in the supporting ring is 6 mm diameter and 6 mm long and the four holes leading to the central region are 2·5 mm, 2 mm, 1·25 mm and 0·95 mm diameter respectively. The tungsten carbide anvils are 1·27 cm diameter with 0·225 mm diameter flats on a 6° taper with shrunk-on steel supporting rings. There is provision for screwed-in brass inserts which serve as piston guides so that the pressure cell between the anvils can be brought accurately into line with the windows.

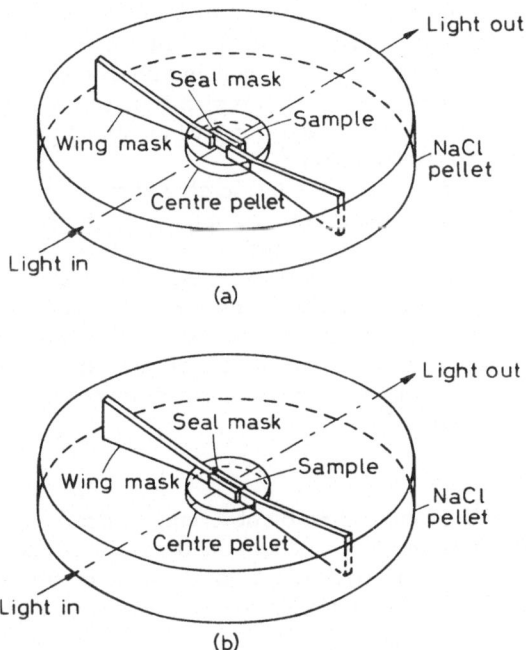

Figure 4.17. Two methods for introduction of sample into the optical cell in Figure *4.16* (after Fitch, Slykehouse and Drickamer[29])

The filling of the windows with sodium chloride is a complicated operation and the procedure used by Drickamer for obtaining the clearest windows is as follows. After thorough cleaning of the holes a single crystal of sodium chloride about 12·5 mm long is shaped roughly to fit into the hole in the supporting ring. It is heated to

77

500°C and inserted through a guide. A tungsten carbide piston fitting the 6 mm hole to less than 0·01 mm is used to work the sodium chloride in by cycling in decreasing steps between 30 kb and atmospheric pressure. After repeating the technique for the second set of windows a 1·5 mm thick crystal is placed in the centre of the cell. After heating to 500°C 30 kb pressure is applied with a flat 1·27 cm diameter piston. Steel plugs should be screwed into the outside of the window holes during this operation and the result is normally a pair of very clear windows. The central section is then removed and the apparatus is ready for inserting the high-pressure cell and supporting plug. Windows made in this way are capable of withstanding a pressure gradient up to 140 kb and provide a spectral range between 0·25 and 10 microns, although the throughput of energy beyond 5 microns from conventional sources may be too low for reliable absorption measurements in some samples to be made. The windows can be used for tens of runs if properly looked after and occasionally repressed.

In using the tapered anvils considerable care has to be exercised to prevent light bypassing the sample. The high pressure cell assembly in *Figure 4.17* is used and the procedure for carrying out a high-pressure experiment is as follows. The stationary piston is inserted into its correct position using the brass guide and a single crystal of salt about 0·30 mm thick fused in position. The masks and sample enclosed in a central pellet are then inserted. The procedure is rather difficult and requires practice before successful cells are made regularly.

Calibration

The cell has been calibrated by observing optically the stretching frequency at 2210 cm^{-1} of CN ions in the sodium chloride lattice and by volume discontinuities at 84·5 kb in silver bromide and at 88 kb in silver chloride. The relation between centre pressure and applied pressure for tapered anvils is strongly dependent on the central thickness (t_c) as the following general expression used by Balchan and Drickamer[30] shows.

$$P_{\text{centre}} = P_{\text{average}}\left[1 + \frac{A}{P^x_{\text{average}}}\exp\left(-Bt_c\right)\right], \quad A, B, x \text{ are constants.}$$

It must be remembered that this equation involves a considerable degree of extrapolation.

In addition it is possible to make electrical resistance calibration in the cell with appropriate insulation of one anvil although the flow of sodium chloride between the anvils smears out transitions too much for accurate work. With extrapolation of the optical shifts a scale up to 140 kb accurate to a few per cent can be found. It is essential to define to about 0·003 mm the initial thickness of the sodium chloride between the high-pressure flats if a reasonable degree of consistency is to be reached.

The use of sodium chloride as both window and pressure medium has proved to be quite successful. It has relatively low sheer strength and acts as a reasonably hydrostatic medium for optical purposes even as low as a few kb pressure. The calibration is usually consistent in itself and its usefulness lies in the possibility of other experimenters referring their pressures to the CN vibration frequency scale rather than to an absolute one which is as yet not well established. In comparison with the cubic anvil optical apparatus which is described in Chapter 5, this cell suffers from a limited spectral range, difficulty in sample assembly and of making simultaneous resistance measurements. However, its pressure range is well over twice that of the cubic device.

High-Temperature Optical Cell

In order to make optical absorption measurements at high temperature simultaneously with high pressure in the cell shown in *Figure 4.16* an internal heater is used[30]. One anvil is insulated in a similar manner to that described for the resistance cell but with pyrophyllite forced into the space around the mica insulation to avoid extrusion of the sodium chloride. The heater is constructed by drilling out 65 gauge holes in the central plug, carefully avoiding the optical path and filling them with powdered graphite. Typical operating conditions for the cell are 400 A at 1·5 V giving a temperature rise of approximately 1 degree per watt above room temperature.

Calibration at Elevated Temperatures

Temperatures may be measured by observing the melting point of substances such as naphthalene. Only a minimum pressure is used to make contact with the heaters. It is possible with careful operation to reproduce the heater position sufficiently accurately to determine the temperature to $\pm 5°$ at 300°C from the number of watts dissipated and a previous calibration. For pressure measurement at high temperatures where redistribution of sodium chloride between the anvils makes the room temperature calibration not applicable the following procedure is used. A spectrum is recorded at room temperature as a function of pressure, the temperature is

79

then raised to some value at a fixed pressure and a second spectrum recorded. After rapid cooling a second room-temperature spectrum is recorded and assuming that the pressure is quenched-in by this process this spectrum should indicate the pressure at the elevated temperature. It is claimed that an accuracy of $\pm$ 3 kb above 30 kb can be obtained.

The cell can be used to 500°C although in the region above 100 kb simultaneously with temperatures above 300°C there is considerable deformation of the anvils.

Low Temperature Optical Cell

Figure 4.18 illustrates a version of the Drickamer optical cell which has been used by Sherman for experiments[31] down to liquid nitrogen temperatures at 50 kb. This is the low-pressure version in which the anvils are really pistons since they are not tapered. The bulk of the cell is constructed from EN 26 steel hardened and tempered to R.C. 50 and cycled between 25°C and −183°C. Larger windows than those used by Drickamer improve the light throughput but it is found necessary to use sapphires as end seals since there is considerable extrusion of sodium chloride above 15 kb at room temperature. The cooling arrangement is such that with suitable

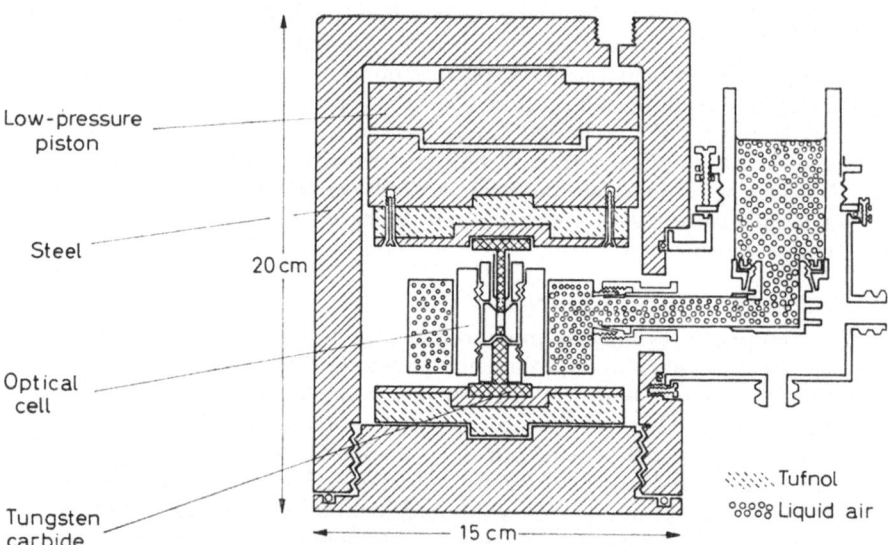

Figure 4.18. Cross section of liquid-air cooled, high pressure absorption cell. The radiation passes through the cell perpendicular to this section (after Sherman[31])

80

insulation between the high-pressure cell and the intensifier piston and evacuation of air the temperature of the outer parts falls by only 15°C when the cell is at -183°C.

The calibration follows that described above with the vibration frequencies of various ions in sodium chloride or caesium chloride lattices. Sapphires are transparent up to 5 microns but other materials may be used instead to widen the spectral range. A loaded cell transmits approximately 25 per cent of incident radiation and spectra with resolution of 1 or 2 cm^{-1} can be obtained.

Experimental Results from the High-Pressure Optical Devices

The series of apparatus which has been described in the preceding sections have proved most useful in the investigation of optical absorption edges in semiconducting and insulating materials. Many III–V and II–VI compounds undergo considerable changes in their electronic band structure when subjected to high pressures which results in one or more crystallographic phase changes. A selection of the results of absorption experiments by Edwards and Drickamer[32] are shown in *Figure 4.19*. These are of first degree importance in interpreting and understanding such phenomena in semiconductors as the Gunn effect and stimulated emission. It is found that phase changes observed optically agree very well with those determined by resistance and x-ray diffraction methods although there may be some small differences in pressure scales. Kluyev[35] has determined the effect of pressure on vibration frequencies in the visible and infra-red using a similar cell to that described above, the main difference being in the use of a tapered sodium chloride window instead of the stepped type of Drickamer.

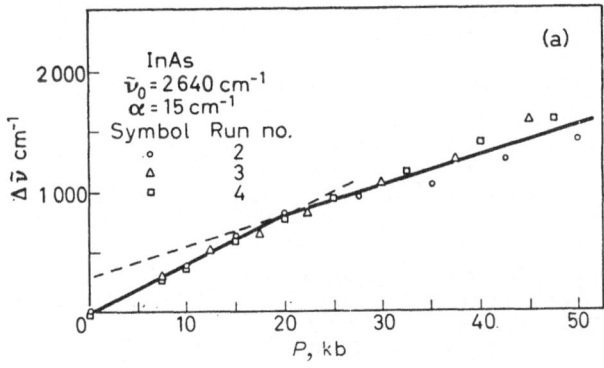

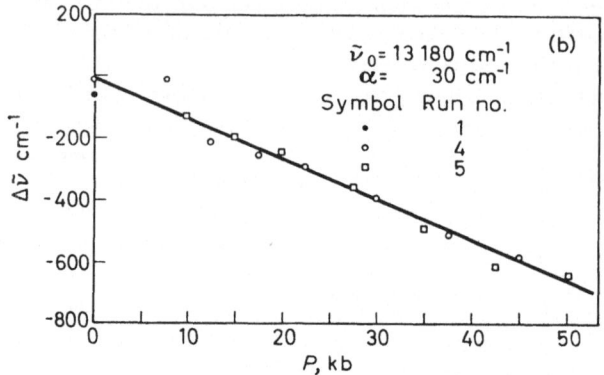

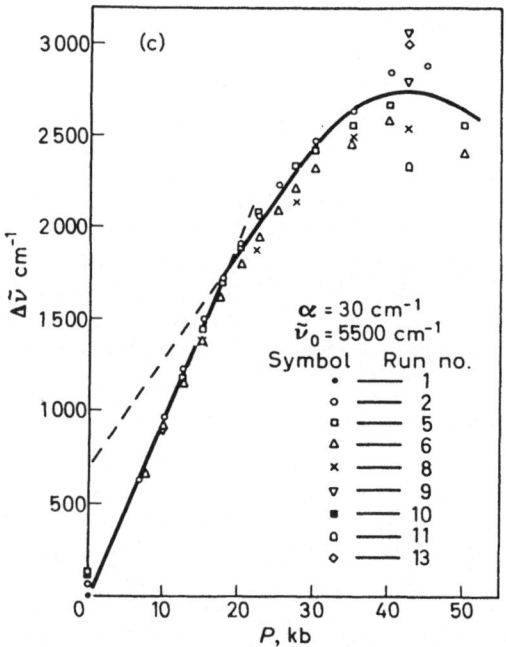

Figure 4.19. Results from optical absorption cell. (a) Shift of InAs absorption edge with pressure. (b) Shift of AlSb absorption edge with pressure. (c) Shift of GaSb absorption edge with pressure (after Edwards and Drickamer[32])

82

High-Pressure Cell for Mössbauer Resonance Measurements

Mössbauer resonance experiments can be used to determine the pressure dependence of internal magnetic fields and electric field gradients at the centre of an atom[33]. The effect of pressure is to introduce extremely small frequency shifts in the gamma ray emission of atoms such as Fe-57. This shift can be determined by using an absorber of the same material and generating a compensating frequency shift by a relative velocity between the source and the absorber. The high-pressure aspect involves the provision of a single exit port for the gamma rays and a pressure medium which is not an absorber. *Figure 4.20* illustrates a pressure cell assembly which can be used with a suitably modified Drickamer cell. Boron with 15 per cent lithium hydride is used between the anvil flats because of its good friction and hence prevention of sample extrusion; the remainder of the plug is a flat disk of lithium hydride 0·37 mm diameter and two pyrophyllite plugs to form supports for the tapered anvils.

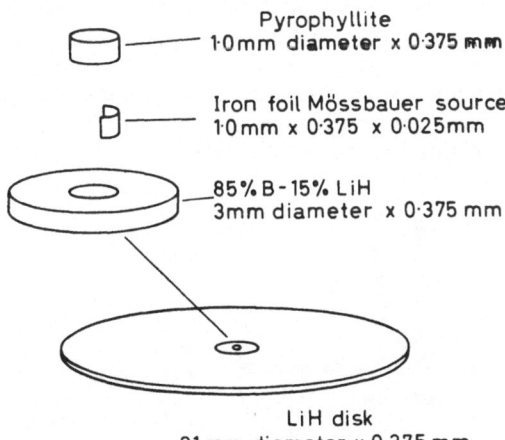

Pyrophyllite
1·0mm diameter x 0·375 mm

Iron foil Mössbauer source
1·0mm x 0·375 x 0·025mm

85% B - 15% LiH
3mm diameter x 0·375 mm

LiH disk
21mm diameter x 0·375 mm

Figure 4.20. Disk of lithium hydride and boron containing Mössbauer source. The details of the centre are shown in enlarged view (after Pipkorn et al.[33])

High-Pressure X-ray Cell

The Drickamer cell which has been described previously can be converted easily to a high-pressure x-ray diffraction apparatus[34] The basic design follows the above optical cells with the press built to fit onto the goniometer of a convenient x-ray set up. The outer steel supporting ring has a 6 mm entrance hole and an exit slit 6 mm deep covering 180° arc and the inner tungsten carbide cylinder has similar ports with 2·4 mm dimensions. The tungsten carbide anvils are 1·27 cm diameter as before with 18° taper and 0·12 mm flats. Support is obtained from pyrophyllite plugs. Between these and coinciding with the centre of the entrance hole and exit slit is a 3·7 mm thick disc of fused LiH. This is a good transmitter of x-rays and contains at its centre an inner disc of 85 per cent boron and 15 per cent LiH together with the sample. LiH/boron mixtures are used because of their greater frictional properties which is important in preventing extrusion from the very high-pressure region. It is necessary to incorporate a satisfactory masking arrangement to prevent x-rays bypassing the sample. This may be done with 0·4 mm platinum wires as shown in *Figure 4.21*. The assembly is first preformed with a pyrophyllite centre which is then removed and replaced by the LiH/boron disc and sample.

Calibrations are made by using a marker material with the sample and measuring the change of lattice dimensions with pressure. Suitable materials are silver and rhodium which are close packed cubic crystals and unlikely to undergo phase changes over a wide range of pressure. (The pressure/volume characteristics for those metals have been measured using shock wave methods by Rice *et al.*[28]) It is essential that calibrations are carried out for each new sample since different samples vary considerably in compressibility and

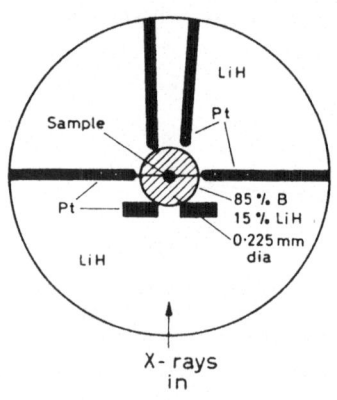

Figure 4.21. Details of LiH disk used for x-ray measurements (after Perez-Albuerne, Forsgren and Drickamer[34])

occupy a large fraction of the central portion of the high pressure cell. Pressures of several hundred kilobars are reached without extrusion of the LiH. The calibration depends considerably on the shock wave data but changes in lattice parameter up to 0·1 per cent can be determined. The limiting factor appears to be the masking of the x-rays due to lensing effects on the flats at the highest pressures.

REFERENCES

[1] Bridgman, P. W. *Proc. Roy. Soc.* 1950, **A203**, 1.
[2] Jackson, J. W. and Waxman, M. *High Pressure Measurement*, p. 39. Eds. A. A. Giardini and E. C. Lloyd. Butterworths, London, 1953.
[3] Christiansen, E. B., Kistler, S. S. and Gogarty, W. B. *Rev. sci. Instrum.* 1961, **32** (7), 775.
[4] Myers, M., Dachille, F. and Roy, R. *High Pressure Measurement*, p. 17. Eds. A. A. Giardini and E. C. Lloyd. Butterworths, London, 1953.
[5] Bridgman, P. W. *Proc. Amer. Acad. Arts Sci.* 1948, **76**, 55.
[6] Kennedy, G. C. and La Mori, P. N. *Progress in Very High Pressure Research*, p. 304. Eds. F. P. Bundy, W. R. Hibberd jr. and H. M. Strong. Wiley, New York, 1961.
[7] Barnett, J. D., Bennion, R. B. and Hall, H. T. *Science*, 1963, **141**, 534.
[8] Montgomery, P. W. Stromberg, H. D., Jura, G. H. and Jura, G. *High Pressure Measurement*, p. 1. Eds. A. A. Giardini and E. C. Lloyd. Butterworths, London, 1963.
[9] Takahashi, T. *High Pressure Measurement*, p. 240. Eds. A. A. Giardini and E. C. Lloyd. Butterworths, London, 1963.
[10] Christiansen, E. B., Kistler, S. S. and Gogarty, W. B. *J. Amer. ceram. Soc.* 1962, **45** (4), 172.
[11] Roy, R. and Cohen, M. H. *Nature, Lond.* 1962, **190**, 798.
[12] Bradley, R. S., Haygarth, J. C. and Munro, D. C. *Trans. Faraday Soc.* 1966, **62**, 2242.
[13] Vereshchagin, L. F. and Brandt, J. V. *Soviet Phys. Dokl.* 1956, **1**, 312.
[14] Kasper, J. S., Hilliard, J. E., Cahn, J. W. and Philips, V. A. *W.A.D.C. Tech. Rep. 59-747.* General Electric Co., 1960.
[15] Jamieson, J. C. and Lawson, A. W. *J. appl. Phys.* 1962, **33**, (3), 776.
[16] Jamieson, J. C. *Physics of Solids at High Pressures*, p. 444. Eds. C. T. Tomizuka and R. M. Emrick. Academic Press, New York, 1965.
[17] Owen, N. B., Smith, P. L., Martin, J. E. and Wright, A. J. *J. Phys. Chem. Solids*, 1963, **24** (12), 1519.
[18] Owen, N. B. and Martin, J. E. *J. sci. Instrum.* 1966, **43**, 197.
[19] Lippincott, E. R. and Duecker, H. C. *Science*, 1964, **144**, 1121.
[20] Martin, J. E. and Smith, P. L. *Brit. J. appl. Phys.* 1965, **16**, 495.
[21] Piermarini, G. J. and Weir, C. E. *J. Res. Nat. Bur. Stand.* 1962, **66A**, 325.
[22] Weir, C. E., Lippincott, E. R., van Valkenberg, A. and Bunting, E. N. *J. Res. Nat. Bur. Stand.* 1959, **63A**, 55.
[23] Connell, N. *Brit. J. appl. Phys.* 1966, **17**, (3), 399.
[24] Van Valkenberg, A. *J. Res. Nat. Bur. Stand.* 1964, **68A**, 97.
[25] Balchan, A. S. and Drickamer, H. G. *Rev. sci. Instrum.* 1961, **32**, 308.
[26] Forsgren, K. F. and Drickamer, H. G. *Rev. sci. Instrum.* 1965, **36**, 1709.
[27] Bridgman, P. W. *Proc. Amer. Acad. Arts Sci.* 1938, **72**, 157; 1951, **79**, 125.

[28] Rice, M. H., McQueen, R. G. and Walsh, J. M. *Solid State Phys.* 1958, **6**, 1.

[29] Fitch, R. A., Slykehouse, T. E. and Drickamer, H. G. *J. opt. Soc. Amer.* 1957, **47**, 1015.

[30] Balchan, A. S. and Drickamer, H. G. *Rev. sci. Instrum.* 1960, **31**, 511.

[31] Sherman, W. F. *J. sci. Instrum.* 1966, **43**, 462.

[32] Edwards, A. L. and Drickamer, H. G. *Phys. Rev.* 1961, **122**, 1149.

[33] Pipkorn, D. N. Edge, C. K., Debrunner, P. de, Pasquali, G., Drickamer, H. G. and Frauenfelder, H. *Phys. Rev.* 1964, **135**, A1604.

[34] Perez-Albuerne, E. A. Forsgren, K. F. and Drickamer, H. G. *Rev. sci. Instrum.* 1964, **35**, 39.

[35] Kluyev, Y. A. *Dokl. Akad. Nauk SSSR*, 1962, **144**, 538.

5

MULTI-ANVIL DEVICES

Multi-anvil devices are extensions in three dimensions of Bridgman's opposed anvils. The latter are not strictly two-dimensional but since the samples are compressed into extremely thin wafers this approximate generalization may be made. By adapting the principle of massive support to apparatus with four or more anvils it has been possible to apply pressures up to 100 kb to sample volumes of several cubic centimetres and upwards. The first person to successfully build such an apparatus was H. T. Hall at Brigham Young University, Utah, who used four anvils with triangular faces to generate pressure in a solid tetrahedron[1]. Obviously other configurations are possible but since the degree of complexity rises rapidly with the number of anvils the only other one to have been developed to any extent is a cubic device with six square-faced anvils[2,3].

The application of the principle of massive support to anvils in a three-dimensional configuration depends primarily on the shape they take in packing closely around the high pressure volume. As will be seen later the anvils which are considered in this chapter are in the form of truncated cones with varying semi-angles. The gain in compressive strength as a function of semi-angle takes the form shown in *Figure 5.1*[4], where experimental results for a number of common materials have been considered. It will be seen that the gain is less the smaller the cone angle. From solid geometry considerations the angles between the slanting faces and the load-bearing surface of the tetrahedral anvils across a given section vary between 71° and 55° and for the cubic anvils they are 45°. Assuming the general data in *Figure 5.1*, the expected gain in compressive strength will be 2·1 and 1·8 respectively. This is obviously a simplification and requires more detailed examination but it is sufficient to give a useful indication. The dotted curve in *Figure 5.1* is interesting since it shows that binding ring support for tungsten carbide anvils is much less effective at semi-angles which are encountered in tetrahedral and cubic devices than in Bridgman anvils. However, it has been shown that binding rings prevent anvils splitting along their axes. Assuming the compressive strength of tungsten carbide to be about 55 kb the upper pressure limits for the two configurations are 120 kb and 100 kb respectively.

87

Anvil faces up to 7·5 cm side dimensions have been used for synthesis work but for most purposes in solid state physics research much smaller sizes are usual. The most convenient method of applying load simultaneously to several anvils depends to some extent on the size of the apparatus. For anvil faces of the order of 1 to 3 cm it is possible to use a wedge reaction method in which the pressure is generated by a uniaxial load. For larger types it is usually more convenient to apply loads individually to each anvil. There is no difference in principle but the former method is generally much cheaper and certainly less complicated. However, in particular laboratory experiments the individual loading technique may be advantageous.

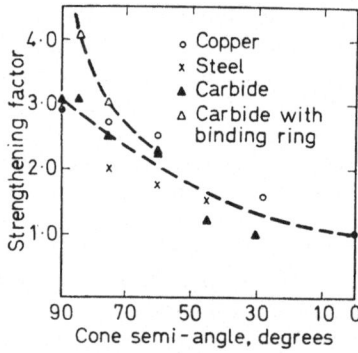

Figure 5.1. The strengthening factor in massive support for conical anvils, as a function of cone semi-angle (after Lees[4])

The two configurations cubic and tetrahedral are basically quite similar. The cube has the advantage of having a straight path through the anvils for optical experiments, better pressure distribution and has two more electrical connections than the tetrahedron. In practice these devices are used up to 70 to 80 kb and temperatures up to 2,000°C.

THE TETRAHEDRAL ANVIL APPARATUS

The first version of this apparatus was built by H. T. Hall and its effectiveness was demonstrated by the synthesis of diamond. (Pressures greater than 70 kb and temperatures greater than 1,000°C.) Other experimenters have used similar apparatus for determining resistivity, volume changes, crystallographic phase changes and many other properties of materials under pressure. The construction and operating conditions have been reviewed very extensively by Lees[4] and the information given in this chapter is meant to indicate the general aspect rather than any critical survey. The method of pressure generation in this apparatus is demonstrated in *Figure 5.2*. Force is applied by anvils simultaneously to the four faces of a solid tetrahedron made from a suitably compressible material slightly larger in edge size than the anvils. Each anvil has three slanting faces at an angle of 71° to the load-bearing triangular face. As load is applied symmetrically to all four anvils the material extrudes into the gaps between them until friction creates a seal as in Bridgman opposed anvils. The success of the operation is due to the fact that if the gasket is formed of a certain type of material, for example pyrophyllite, it remains compressible up to extremely high loads, thus ensuring an increase of pressure inside the tetrahedron. If this were not so any increase of load would be simply taken on the solid gasket.

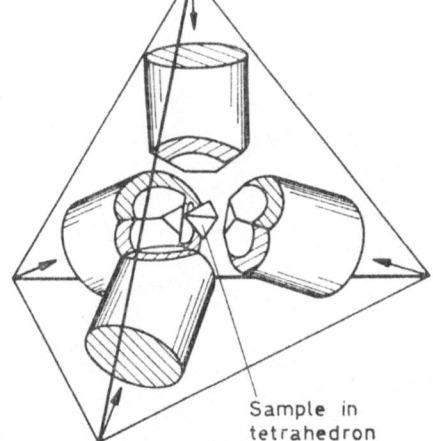

Figure 5.2. Tetrahedral anvil assembly

Sample in tetrahedron

The case considered first is the National Bureau of Standards adaptation since this is more convenient for laboratory use with small-scale apparatus[5]. The load on three of the four anvils is generated by a wedge reaction to the applied force on the fourth anvil. This is illustrated in *Figure 5.3*. By this means uniform loading on all the anvils is obtained from an uniaxial force in a conventional press.

The construction of a typical anvil is shown in *Figure 5.4*. This follows the usual method of using tungsten carbide inserts with hardened steel supporting rings although the effectiveness of the latter is much reduced in this case for the reasons given in *Figure 5.1*. There is a variety of constructions for the inserts and binding rings

89

but only with those most commonly used will be described. Typically, the tungsten carbide insert has a 2 cm side triangular face ground on it and has overall dimensions 4 cm diameter and 4 cm long. A steel binding ring of EN 26 hardened and tempered to R.C. 40–42 is heat shrunk on to it with a 0·04 cm interference fit. Alternatively tungsten carbide may be replaced by Stag Major steel (R.C. 63–65) or the whole unit may be of monoblock construction of Edgar Allen Double 6 tool steel hardened and tempered to R.C. 62. Three slanting faces at angles of 71° to the triangular face are ground on the anvils so that the four pack into a solid unit around the central tetrahedron (*Figure 5.2*). In practice only the

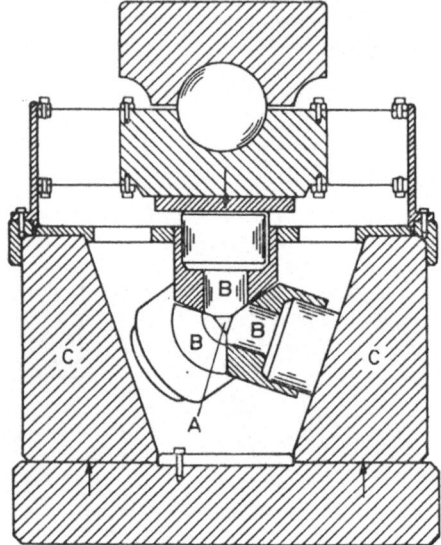

Figure 5.3. A, test sample in pyrophyl-lite tetrahedron; B, loading anvils; C, cone wedge

first 3 mm from the triangular side is at 71°, the remainder being at 1° greater angle so that as the anvils come together under load the chance of accidental contact and hence electrical shorting in regions beyond the gasketting area is reduced to a minimum. The back of the steel supporting ring is counterbored such that about 0·05 mm of the tungsten carbide insert is proud. The anvil is completed by a steel backing block (R.C. 50) in which is placed a pad of tungsten carbide as shown. This acts as an extra support for the anvil face and the 0·05 mm extension of the insert is to ensure that load is taken on the pad. For three of the four anvils the rear face of the backing block is shaped to fit the inside of a cone of 18½° included

semi-angle such that a reaction normal to the interface passes 1°
away from the perpendicular to the triangular face of the head
of the anvil. (See below for explanation of the 1° displacement.) A
special guide block is used to machine the rear faces at the correct
angles. The fourth anvil has a plane rear surface (*Figure 5.4*) onto
which a uniaxial load is applied. In order to facilitate the replacing
of damaged tungsten carbide inserts the backing block and upper
part of the anvil are connected by means of a pair of locating keys.
This helps to keep the correct orientation between them. The whole
unit is held together by an outer keeper tube of mild steel which is
clamped on by Allen screws. A hollow ring may be pressed onto
the keeper ring so that heating or cooling liquids can be pumped
around the anvil. Also a heavy brass terminal for electrical purposes
is usually soldered to it.

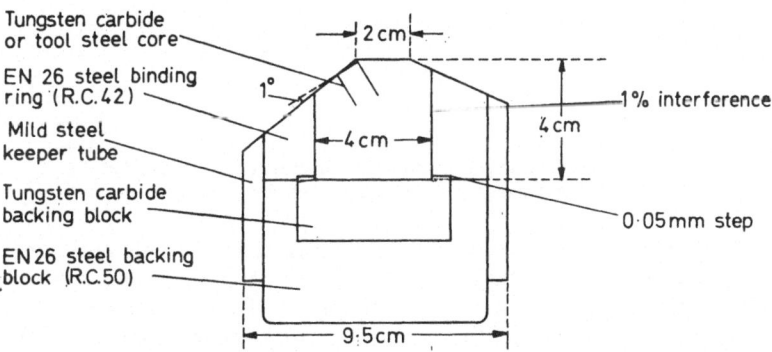

Figure 5.4. Tetrahedral anvil construction (after King[7,8])

A containing wedge cone is of the order shown in *Figure 5.3*. It
may be constructed of EN 25 steel (R.C. 40) and 23 cm maximum
internal diameter, 20 cm deep and 35 cm outside diameter. The
included semi-angle of the cone is $18\frac{1}{2}°$ which is 1° less than the
theoretical geometrical value. The reason for this is the finite friction
between the backs of the anvils and the inner surface of the cone.
The equivalent friction angle is approximately 1° and it will be
seen from *Figure 5.5* that if the backs of the three anvils fitting into
the cone are at a 1° angle to the plane of the triangular faces the
net force of the reaction and friction forces is then perpendicular to
these faces. This is most important for the formation of uniform
gaskets. Sheets of mylar and teflon 0·01 mm thick are placed
between the anvils and the cone to improve lubrication and to
electrically insulate each anvil from the other. The upper anvil

should have its back face and load bearing face parallel to a very high degree of accuracy and the cone must be aligned with the axis of the press. These conditions may be assisted if a ball seating for the loading pad is used (*Figure 5.3*).

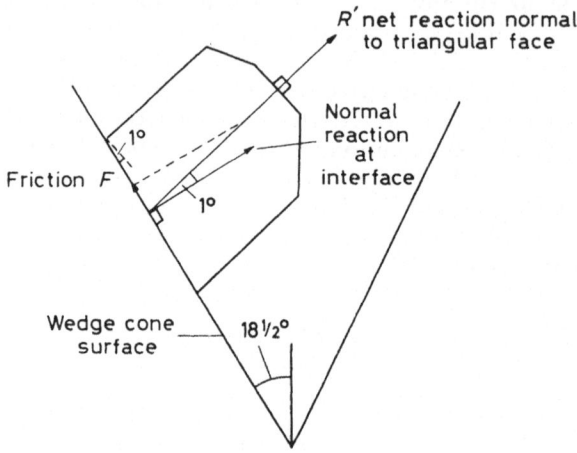

Figure 5.5. Friction and normal forces during wedge-reaction mechanism (angles exaggerated for clarity)

A 2 cm edge size tetrahedral apparatus, that is anvils and cone, weighs approximately 50 kg and normally requires a crane to lift it. This really limits the upper size of the anvil faces to about 4 cm since the cone would then become too massive for easy manipulation. Apparatus much smaller than this is correspondingly easier to handle but as will be seen later the pressure conditions inside the tetrahedron tend to be such that the fraction of the volume over which pressure is reasonably hydrostatic is only a few per cent of the total, thus acting against small sizes. If a large press of the Amsler or Farnell type is available (Appendix B), the cone may be permanently mounted on a small railway and wheeled into the press as desired.

Operation

The three anvils with curved backs are placed in position with mylar and teflon sheets held by adhesive tape between them and the cone surface. Three small buttons of teflon 2 mm thick are placed between coincident slanting faces in order to keep them a sufficient distance apart to begin with so that the initially oversize pyrophyllite tetrahedron can be put symmetrically in position.

For 2 cm anvils a tetrahedron of between 2·2 and 2·5 cm side is used. A vertical load is applied to the plane face of the fourth anvil and some is taken initially by the teflon buttons until the tetrahedron is being deformed. The gasketting regions on the anvils (approximately 3 mm wide) should be coated with jeweller's rouge to enhance friction. The initial size of the pyrophyllite tetrahedron depends on a number of factors and the optimum will usually be determined by the experimenter himself. Generally speaking the greater the oversize the bigger the gaskets and hence the greater pressure range because of the increased compressibility but the efficiency generated in terms of kilobars per load is usually less[2]. At the end of an experiment the anvils are simply lifted out of the cone. One of the important advantages of the tetrahedral anvil apparatus is the ease with which samples and pressure media are inserted and removed compared with other methods.

The life of the anvils depends on the upper pressure limit and temperatures used and the care with which the apparatus is set up. Although the tungsten carbide inserts may crack quite early on it is possible for them to be used until this is so bad that the gaskets do not form properly. Tungsten carbide inserts will probably last for twenty or more runs to 50 to 70 kb and cost about fifteen pounds sterling each to replace. If temperatures well above 500°C are used the lifetime is reduced very considerably.

Pressure Distribution and Calibration

By placing thin wire meshes in a tetrahedron it is possible to study the distribution of pressure throughout its volume. It has been found that approximately homogeneous conditions exist in only a few per cent of the total volume, the maximum deviation being in the gasketting regions as expected[4]. This is also confirmed by noting the phase transitions in bismuth wires placed in different parts of the tetrahedron. For this reason accurate measurements are confined to samples less than a few mm in the above size tetrahedral anvil apparatus.

Calibrations are made conveniently with the usual electrical resistance discontinuities in bismuth, thallium and barium. A suitable cell assembly in an 2·2 cm pyrophyllite tetrahedron is shown in *Figure 5.6* where a thin wire sample about 0·25 mm diameter is placed inside a silver chloride sleeve 4 mm diameter and 8 mm long and current and voltage contacts made to the four anvils with thin nickel tabs. The presence of silver chloride (or indium) in the centre of the tetrahedron improves the uniformity of pressure. The resistance of the anvil is usually very much less than that of the

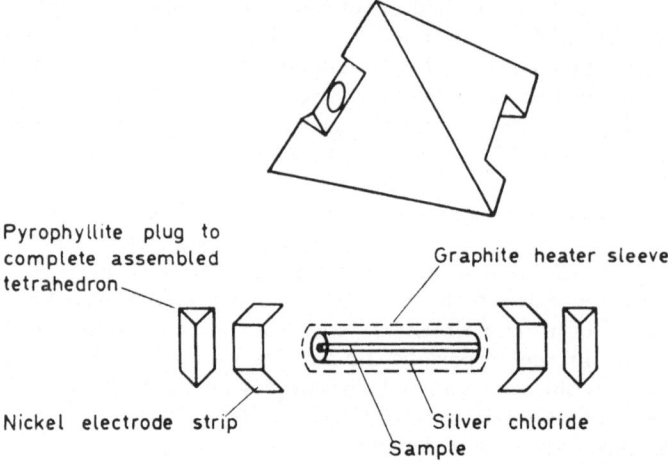

Pyrophyllite plug to
complete assembled
tetrahedron

Graphite heater sleeve

Nickel electrode strip

Silver chloride
Sample

Figure 5.6. Edge-to-edge pyrophyllite tetrahedron sample assembly
(Note. If the heater tube is used a different set of electrode connections
are made)

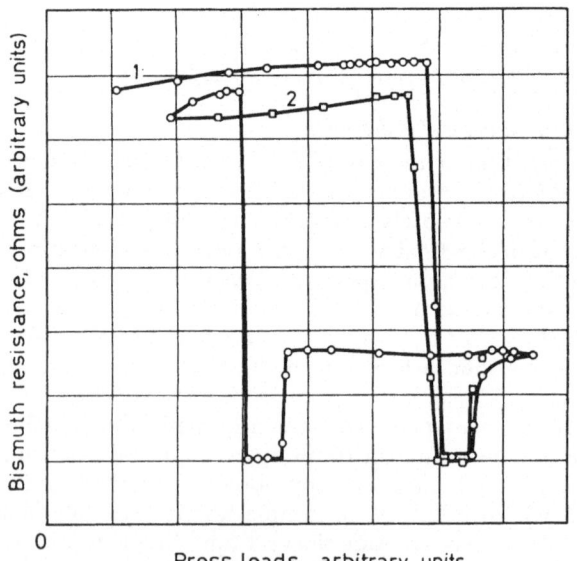

Bismuth resistance, ohms (arbitrary units)

0

Press leads, arbitrary units

Figure 5.7. I–II and II–III transition in bismuth showing
hysteresis on unloading (after Bundy, F. P., A.S.M.E.
60-WA-178, 1961, p. 1)

samples and only two anvils are needed for connection to the electrical circuit. The samples should be coated in epoxy resin as before (Chapter 4) to prevent corrosion by silver chloride.

The compressibility of the sample cell is an important factor in determining a load/pressure relation and care must be taken to define a particular assembly before using calibration curves. Although the sample and sleeve comprise only about 10 per cent of the total volume, replacing the silver chloride by a material of different compressibility, for example polythene, may change the load to reach a required pressure by more than 10 per cent[4].

Large hysteresis is normally experienced in the loading and unloading pressure cycle as shown in the example of the bismuth transition in *Figure 5.7*. This is due mainly to irreversible changes during gasket formation and it is normal to define a transition by the point at which a discontinuity is just commencing.

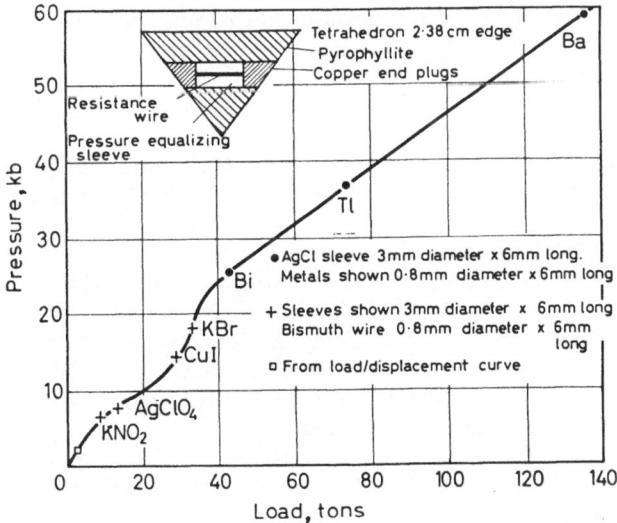

Figure 5.8. Load/pressure curve for 1·9 cm edge N.B.S. type tetrahedral anvil apparatus with (inset) tetrahedron assembly (after Lees[6])

For calibrations in the range above the bismuth I–II point extrapolations between the thallium (37 kb), barium (59 kb) and upper bismuth (89 kb*) transitions are reasonably good to about 3–5 per cent. However below 26 kb there are very few transitions

* See footnote on page 4.

which can be determined by an electrical resistance discontinuity method (only cerium between 7 and 9 kb and this is too unreliable and difficult for use). It is not correct to simply extrapolate straight to the origin since distortions from a smooth curve will probably occur at loads coinciding with the formation of the gaskets. There are a number of compounds which have large volume discontinuities in this pressure region and which have been determined using piston and cylinder apparatus. These include KNO_2, $AgClO_4$, $AgNO_3$, CuI and KBr. A method has been developed at the Standard Telecommunications Laboratory in which these are used as sleeve materials for experiments with bismuth wires or any other material with a smooth resistance versus pressure curve below 25 kilobars[6]. The resistance versus load characteristics of these cell assemblies show small discontinuities in slope at the appropriate volume transitions and a curve of the form shown in *Figure 5.8* may be obtained. As has already been stated above the compressibility of the sleeve material is important and it must be corrected for by experiments with silver chloride. The accuracy of this calibration is from 3 to 5 per cent.

Efficiency

The efficiency is measured in terms of load required to generate a given pressure and can be specified only for a well defined system.

Effect of Anvil Materials[7]

A load versus pressure relation using the three different forms of anvils described above is shown in *Figure 5.9*. The tungsten carbide variety are the most efficient and there is very little difference between the multi-steel and monoblock types. At the highest loads the steel anvils probably deform more than the tungsten carbide ones and less pressure is generated[7].

It has been found that changing the anvil size has practically no effect on the efficiency other than by the area multiplication factor[2].

Effect of the Gasket Material

Beside pyrophyllite, talc, boron nitride, and mixtures of lithium hydride and boron and other substances have been used as gasketting materials. With regard to pressure generation efficiency and ease of gasket formation, pyrophyllite is superior to any of these[8]. A new material developed at the Standard Telecommunications Laboratory[4] consists of a 50–60 per cent mixture by volume of magnesium oxide (200 grade) and epoxy resin (Araldite MY 704). The tetrahedra are formed in moulds and cured up to 180°C. They have been found to be up to 25 per cent more efficient than pyrophyllite and raise the ultimate pressure limit from 70 to 100 kb. The material suffers from the disadvantage of not having good electrical and mechanical properties at high temperatures.

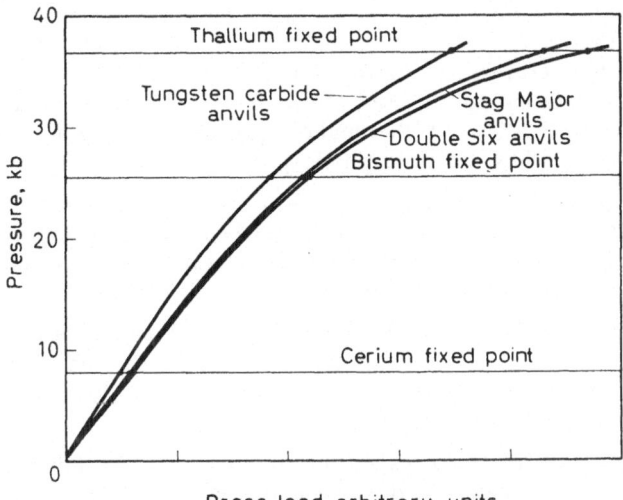

Figure 5.9. Pressure calibration for tetrahedral anvils of different construction (after King[7])

Pyrophyllite can be subjected to varying degrees of heating and compression treatment before use in the high-pressure apparatus but the effects of this on pressure generation efficiency are to some degree indeterminate.

Small discs of mild steel are sometimes placed in the faces of the tetrahedron in order to reduce the overall compressibility. It is possible to improve the efficiency by more than 20 per cent by this means but the uniformity of pressure distribution suffers considerably.

Heating

The great advantage of tetrahedral anvil apparatus is that comparatively large sample volumes may be heated to temperatures up to 2,000°C. The system is reasonably thermally efficient since although the anvils provide a large heat sink the thermal conductivity of pyrophyllite is low enough to maintain a high internal

97

temperature. It is usually found that even at the highest temperatures the anvil faces are at only one or two hundred degrees and
hence cooling is unneccessary. Heaters may be made in the form of
either hollow cylinders (about 0·2 mm wall thickness) of graphite or
a refractory metal like tantalum or thin metal strips. As a general
guide the generation of heat in the above size of apparatus is
approximately 1° per watt up to 1,000°C. Uniformity of pressure
at high temperatures is improved with boron nitride sleeves since
silver chloride cannot be used above 400°C. Pyrophyllite remains
solid up to 2,500°C at the highest pressures (100 kb) although
melting at 1,500°C at atmospheric pressure.

Edge to edge assembly (Figure 5.6)—In this arrangement the heater
tube has maximum possible length and hence highest temperatures
are reached using it. The disadvantage is that there is considerable
interaction with the gasketting regions.

Face to vertex (Figure 5.10a)—This is a method which has been used
for synthesis work. The graphite tube breaks through the tetrahedron at all three faces at the vertex and the three anvils are
connected together as one heater terminal.

Face to face (Figure 5.10b)—This method has the advantage of
avoiding gasket regions and is reasonably efficient thermally.

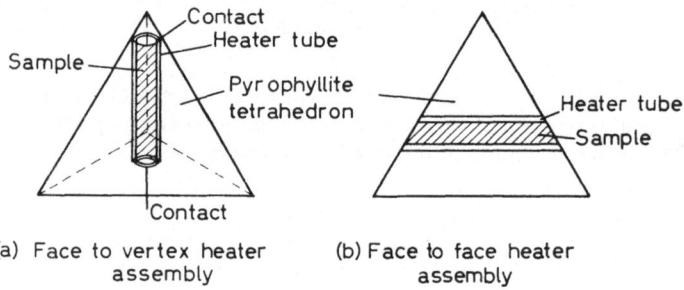

(a) Face to vertex heater
assembly

(b) Face to face heater
assembly

Figure 5.10. Heater assemblies

Temperatures are measured by inserting thermocouples at the
appropriate points in the tetrahedron. The leads are arranged to
pass through the gaskets which sometimes creates difficulties with
pinching-off due to pressure inhomogeneites during gasket formation.
A useful thermocouple is the Philips thermocoax type which consists of chromel and alumel wires embedded in magnesium oxide
inside a stainless steel sheath 0·25 mm o.d. (A calibration curve

under one atmosphere conditions can be obtained from the suppliers.) The effect of pressure on thermocouples has been investigated up to 80 kb and 1,500°C and the reader is referred to Chapter 1 for details.

The effect of high temperature on pressure calibration has been studied using the melting point of zinc and aluminium as functions of pressure[9]. In the region up to 40 kb there are data from piston and cylinder and hydrostatic experiments[10,11]. The results indicate that at 40 kb and temperatures around 1,000°C the load/pressure curve changes by about 10 per cent (see Chapter 1).

OTHER EXPERIMENTS IN THE TETRAHEDRAL ANVIL APPARATUS

Synthesis

Diamond—Diamonds have been synthesized in this type of apparatus by many experimenters. An assembly used by Pugh and Lees[12] in a 2 cm anvil apparatus is shown in *Figure 5.11*. The pressures and temperatures are 60 kb and 2,000°C.

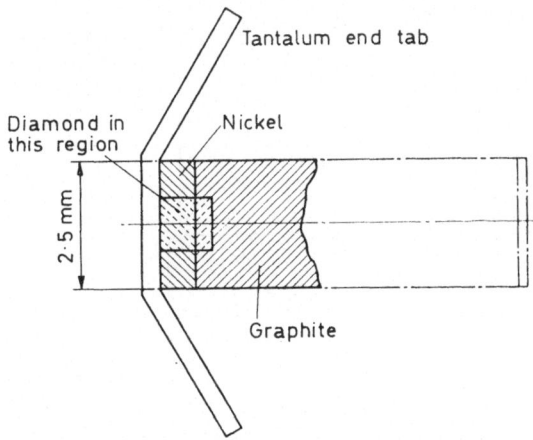

Figure 5.11. High-pressure cell used for diamond synthesis in a tetrahedral anvil apparatus (after Pugh and Lees[12])

Polymeric carbon disulphide—This synthesis has been performed at the National Physical Laboratory using the assembly in *Figure 5.12*[13]. A teflon bottle holding liquid CS_2 is placed in the tetrahedron and solidified by cooling the anvils down with circulating liquid nitrogen. A small load is applied to make a good seal and after but only with those most commonly used will be described. Typically, the tungsten carbide insert has a 2 cm side triangular face ground on it and has overall dimensions 4 cm diameter and 4 cm long. A steel binding ring of EN 26 hardened and tempered to R.C. 40–42 is heat shrunk on to it with a 0·04 cm interference fit. Alternatively tungsten carbide may be replaced by Stag Major steel (R.C. 63–65) or the whole unit may be of monoblock construction of Edgar Allen Double 6 tool steel hardened and tempered to R.C. 62. Three slanting faces at angles of 71° to the triangular face are ground on the anvils so that the four pack into a solid unit around the central tetrahedron (*Figure 5.2*). In practice only the

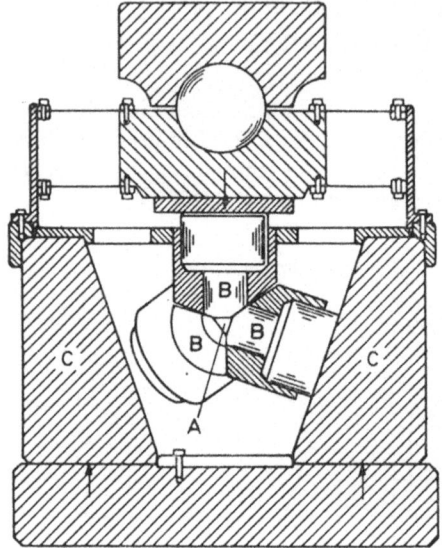

Figure 5.3. A, test sample in pyrophyllite tetrahedron; B, loading anvils; C, cone wedge

first 3 mm from the triangular side is at 71°, the remainder being at 1° greater angle so that as the anvils come together under load the chance of accidental contact and hence electrical shorting in regions beyond the gasketting area is reduced to a minimum. The back of the steel supporting ring is counterbored such that about 0·05 mm of the tungsten carbide insert is proud. The anvil is completed by a steel backing block (R.C. 50) in which is placed a pad of tungsten carbide as shown. This acts as an extra support for the anvil face and the 0·05 mm extension of the insert is to ensure that load is taken on the pad. For three of the four anvils the rear face of the backing block is shaped to fit the inside of a cone of $18\frac{1}{2}°$ included

90

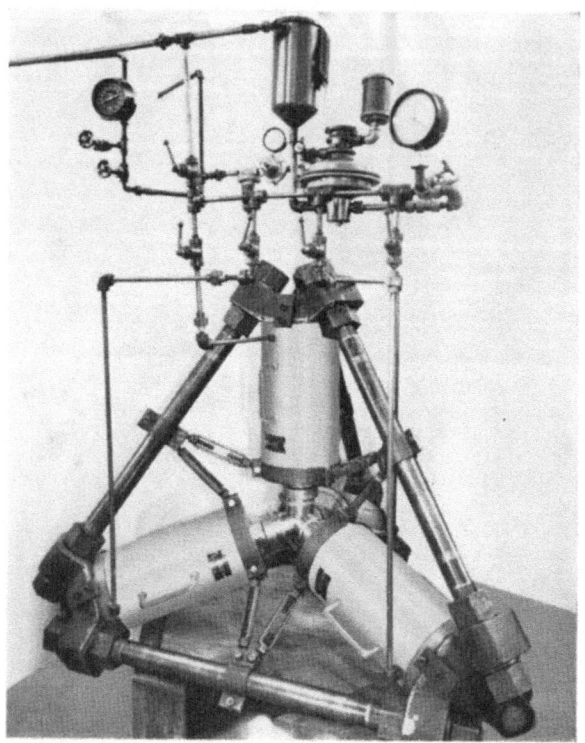

Figure 5.13. Individual ram tetrahedral anvil apparatus
after Hall

(*To face p. 101*)

More sophisticated analyses are possible and the reader is referred to the work of Claussen[14] and Blum and Deaton[15].

INDIVIDUAL RAM TETRAHEDRAL APPARATUS

In the tetrahedral anvil apparatus used by Hall[1], a system of tie rods and individual hydraulic rams applies load to the tetrahedron, and although the construction is more complicated it has enough advantages to warrant a description here. It is especially useful where large scale uniaxial presses capable of up to several hundred tons loading are not readily available. It is very often more convenient to use a supply of oil at fairly high pressures which may be placed some distance from the experimental area. The absence of any surrounding wedge cones makes the apparatus attractive for combination with x-ray diffraction equipment.

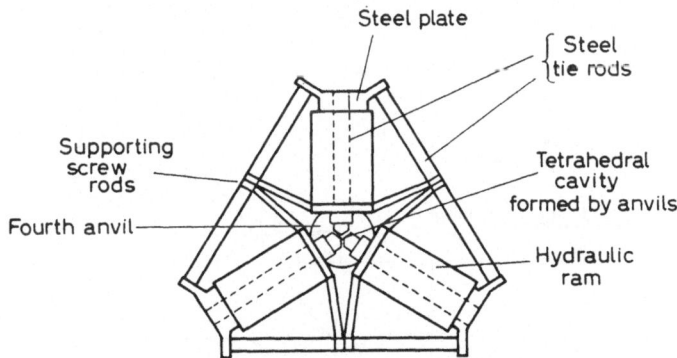

Figure 5.14. Individual ram tetrahedral apparatus

The construction is shown in *Figures 5.13* to *5.15*. The design of the anvils is very similar to those described above, that is with tungsten carbide heads and backing pad and a shrunk-on supporting ring of hardened steel which is in this case shaped on its rear face so that a steel block and the piston from an hydraulic ram fit snugly into it. Provision for electrical insulation must be made here. Each of the four anvils is driven by a cylindrical pressure ram in which oil up to 1 kb is pumped. The high pressure piping is arranged so that pressure can be applied separately to any of the anvils or to all four simultaneously. At the backs of each of the hydraulic rams there is a large steel plate on to which are bolted symmetrically three tie rods forming the symmetrical polygon as shown. The

101

tie rods are of steel and up to 7·5 cm in diameter. Screw rods positioned half-way along each tie rod support the cylindrical rams. It is important that the alignment of the support system should reach a very high degree of precision since any slight deviation causes damage to the anvil tips when load is applied.

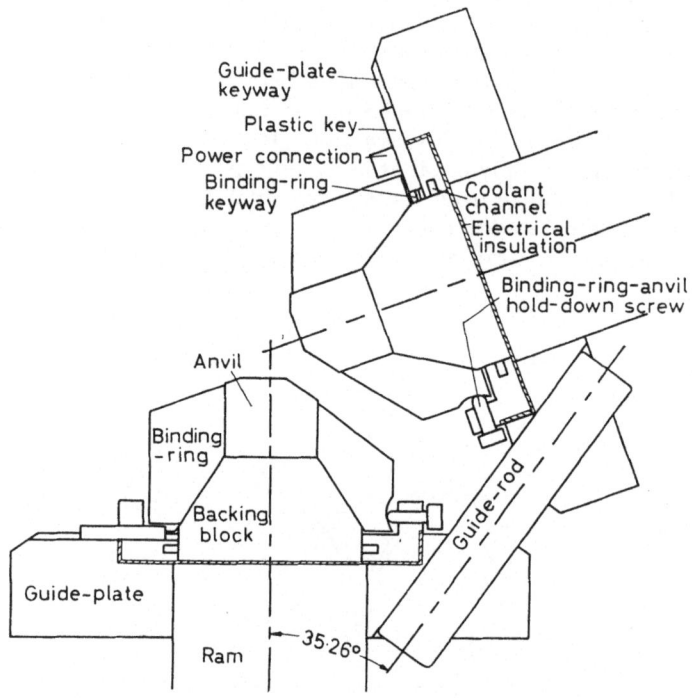

Figure 5.15. Plane section through axes of two hydraulic rams of a tetrahedral press showing anvil guide device (after Hall[16])

With this type of construction pressure is applied to a pyrophyllite tetrahedron by first increasing the load in each ram successively by small increments until a reasonable degree of symmetry has been obtained and finally to all four rams simultaneously up to the required value. In order to eliminate the need for this somewhat tedious and skilful procedure Hall has designed a guide system which ensures accurate alignment[16]. This is illustrated in *Figure 5.15.* A guide plate (steel) is fitted to each anvil supporting ring, again making arrangements for electrical insulation. These are drilled out with three symmetrical holes making angles of 35° 26' with the tie

Figure 5.16. Tetrahedral press at the Monroeville, Pa., plant of the U.S. Steel Corporation (Zeitlin[17])

rods. Six guide rods are inserted and interconnect the four guide plates. When pressure is applied simultaneously to all four anvils they must move together symmetrically within the accuracy of the machining of the guide system.

Although tetrahedral anvil apparatus with tie rods have been produced commercially on a miniaturized scale they have proved too costly to compete with the wedge action type. At the other end of the scale a very much larger version has been produced for commercial synthesis by Barogenics Incorporated. For this purpose large sample volumes are required and anvil edges up to 10 cm in size are used. With oil pressure in the rams at 1 kb the intensifier ratio has to be of the order of 100 to 1 to produce the required loads at the anvil faces. As will be seen in *Figure 5.16* the tie rod system is replaced by large steel plates interlocked with linch pins. The advantage is that whereas in the Hall tie rod system all the force is taken on a single cross-section of the rod, here it is distributed in the hinge along the linch pin. It is claimed that the method reduces some of the deflecting forces which occur in the tie rod construction and therefore improves the alignment[17]. The scale of the apparatus makes it applicable only for commercial use except in unusual circumstances.

TETRAHEDRAL ANVIL X-RAY DIFFRACTION APPARATUS

The tie rod tetrahedral anvil apparatus has been adapted for x-ray diffraction measurements under pressure by Barnett and Hall[18]. It has two main advantages over the Bridgman anvil type of x-ray squeezer and piston and cylinder devices constructed from beryllium or diamond etc. Firstly, the pressure is developed three-dimensionally as opposed to the approximately two-dimensional Bridgman anvils and although the volume of hydrostatic pressure region may be only a few per cent of the total tetrahedron size, it is quite large enough for x-ray diffraction measurements. Secondly, pressures up to 100 kb and temperatures of 1,000°C are possible in materials under near hydrostatic condition. These limits are well beyond those possible in beryllium and diamond cylinders especially in the latter case where graphitization begins above 600°C. The apparatus is illustrated in Figures *5.17* and *5.18*. The accuracy of the machining is to a very high order consistent with x-ray diffractometry techniques.

The x-ray tube is mounted in one of two positions. In *Figure 5.18A*, the entry of the beam is through a gasket and the faces of the anvils

103

are bevelled by 4° to increase the solid angle of entry and exit and hence radiation throughput. The diffraction pattern is observable over a 2θ angle of 0 to 155° and by using slits and scintillation counters at the three other gasket positions, three diffraction patterns of the same system are obtained from which accurate measurements may be made. The diffracted x-ray beam travels through a considerable thickness of the pressure medium and hence a detection system of high discrimination is necessary. In the configuration of *Figure 5.18B* one ram and anvil is bored out to take the x-ray entry beam. At the high pressure face the hole is 0·7 mm diameter and widens out to a small cone shape carrying a beryllium plug. The latter is used to prevent extrusion of the gasketting material. In this configuration the 2θ angle is −55° to +55° which means greater

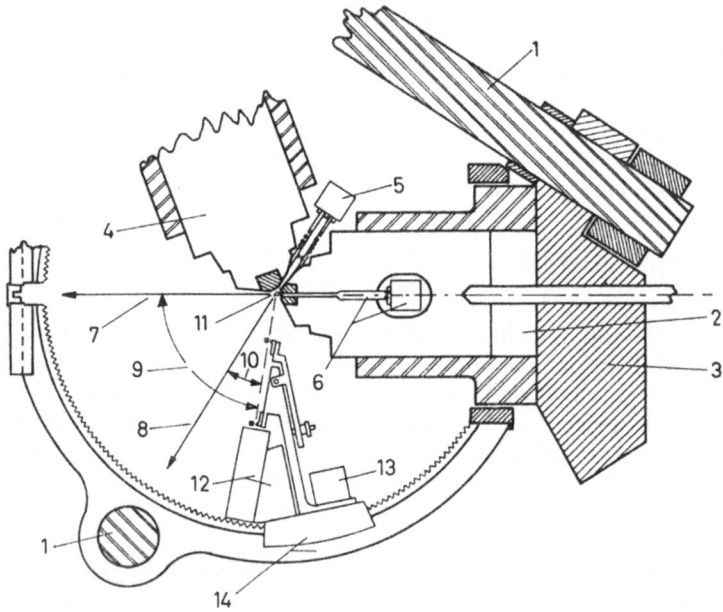

Figure 5.17. Cross-section of x-ray tetrahedral anvil apparatus showing the x-ray tube mounted in each position and the detector scanning mechanism as they relate to the high pressure system. 1, tiebar; 2, hydraulic oil; 3, ram base; 4, piston assembly; 5, x-ray tube and collimator in position B; 6, x-ray tube and collimator in position A; 7, undeviated x-ray beam, position B; 8, indeviated x-ray beam, position B; 9, diffraction angle (2θ), position A; 10, diffraction angle (2θ), position B; 11, sample; 12, scintillation counter and pre-amp; 13, scanning motor; 14, scanning carriage and track (after Barnett and Hall[18])

accuracy for the estimation of lattice parameters but less certainty in structure determinations since fewer lines are observable. It has been found that the perforated anvil may be used up to 70 kb without noticeable weakening. The diffractometer components are rigidly attached to the high-pressure tie rods. The accuracy with which lattice parameters can be determined is approximately to $\pm 0 \cdot 1$ per cent using the arrangement of *Figure 5.18B*.

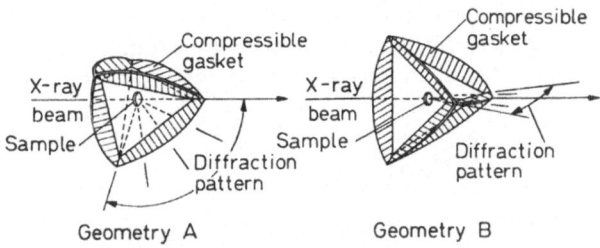

Geometry A   Geometry B

Figure 5.18. Tetrahedral sample chamber showing the two possible x-ray geometries using the compressible gasket as the exit pupil (after Barnett and Hall[18])

The question of a suitable pressure medium and gasketting material presents the usual problems in x-ray diffraction. Both lithium hydride and boron are reasonably transparent to x-rays but lithium hydride extrudes too readily and has little friction and boron is the reverse of this. For normal temperatures (up to a few hundred degrees) a 50/50 mixture of these two substances has been found to be quite suitable. The sample whose x-ray diffraction pattern is to be determined is placed in the centre of a tetrahedron sandwiched between small pieces of polythene (below 400°C) or hexagonal boron nitride (greater than 400°C). Heating may be incorporated by means of graphite or metal sleeves, careful positioning being necessary only in the case of the latter since graphite is reasonably transparent to x-rays. For temperatures above 500°C Barnett and Hall have found that tetrahedra constructed from boron-filled plastic material are very suitable. (The plastic resin they used is obtained from the Hooker Chemical Company and is known as Durez 14684. Fifty per cent by weight mixtures are used.)

As an illustration of the capabilities of this apparatus the following experiments are quoted. In one investigation the diffraction pattern of potassium chloride was obtained at a series of pressures up to 20 kb each one taking thirty minutes exposure time. In the case of materials of higher Z number, for example barium, this may be

105

increased to twenty hours. This illustrates the severe requirements on the mechanical stability of the apparatus.

Calibration

The x-ray tetrahedral device is probably the most versatile apparatus for pressure calibration. The lattice parameter of sodium chloride which has a cubic structure has been measured as a function of pressure and temperature and it is suggested that a new pressure scale based on the equation of state of sodium chloride should replace the one derived from the electrical resistance discontinuities in a number of chosen materials[19] (see page 9). *Figure 5.19* shows

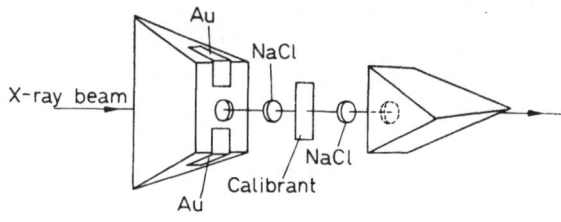

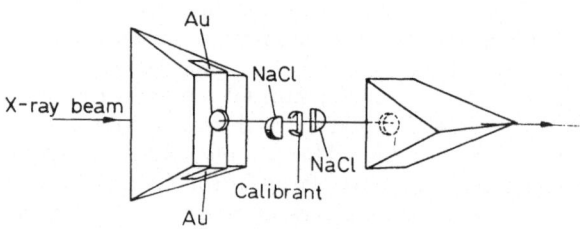

Figure 5.19. Construction of calibration samples showing location of calibrant, NaCl and x-ray beam (after Jeffery et al.[19])

an experimental arrangement in which the lattice parameter of sodium chloride, the lattice parameters of a calibrant metal and the resistance of the calibrant metal can be determined simultaneously. In this way the two components are kept in such intimate contact that there should be no difference in pressure.

The use of sodium chloride as a pressure indicator over the range obtainable in this apparatus has begun to be accepted only recently because of some unconfirmed earlier reports of a pressure transition at around 20 kb which would of course invalidate the use of

Decker's[20] equation of state to yield pressure as a function of lattice parameter. With the above accuracy for lattice parameter determination pressure differences of less than 1 kb may be measured. The reader is referred to Chapter 1 for further details on calibration.

CUBIC ANVIL APPARATUS

The second multi-anvil apparatus device to be used to any great extent is in the form of a cubic configuration in which six anvils with square faces apply pressure to a cube of pyrophyllite or similar substance. As in the case of the tetrahedron there are two versions, one for apparatus on a large scale with individual rams applying load to the anvils and the other an adaptation of the above wedge cone apparatus. Most normal laboratory experiments are concerned with the second type because of its comparative simplicity and cheapness. The device described here is one that has been developed at the National Physical Laboratory and follows basically the National Bureau of Standards design[2]. It has been adapted for use as an optical absorption cell for frequencies ranging from ultra-violet to the submillimetre region[21,24].

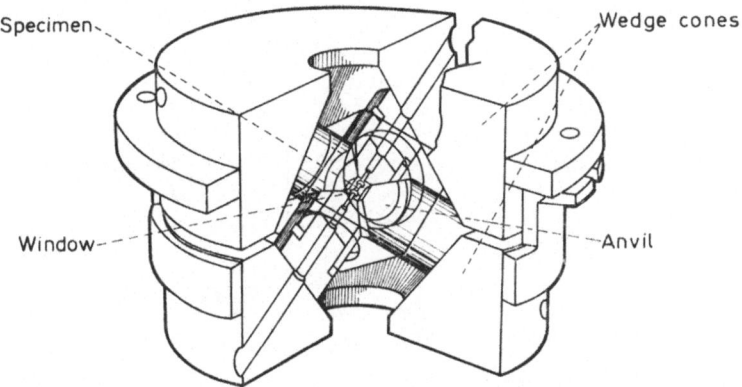

Figure 5.20. High pressure cube-optical apparatus (after Owen[21])

In principle the six anvil cube apparatus is very similar to the tetrahedral device. The construction is shown in *Figures 5.20, 5.21* and *5.22*. The maximum pressure obtainable in this version is about 60 kb with temperatures up to 1,500°C. The anvils consist of an anvil head of tungsten carbide 2·5 cm diameter and 2·5 cm long with a square face 6 mm side dimension and an EN 26 steel ring

(R.C. 42) heat shrunk on to it with an interference fit of 0·02 mm. The head and supporting ring are ground with four slanting surfaces at 45° to the square face so that the anvils fit together into a solid unit. As in the case of the tetrahedral anvils the slanted faces are relieved to give an angular clearance of 2° up to within 0·25 cm of the high-pressure bearing square face in order to prevent touching in regions away from the gasket. The tungsten carbide insert is backed by a piece of EN 26 steel (R.C. 50) whose rear face is shaped to match the tapered wedge cone. The steel ring locates with the rear block through a pair of keys and is locked in position with small Allen screws. For use at lower pressures the anvil can be made from a single piece of Stag Major high-speed tool steel hardened and tempered to R.C. 62. The anvils are fitted with terminals for electrical contacts and with pressed-on hollow rings for circulating cooling fluids.

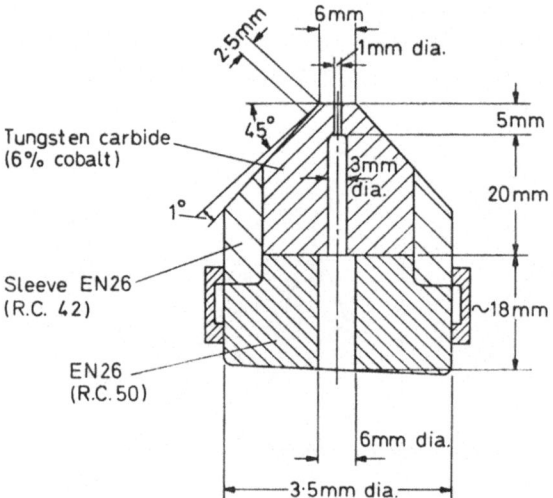

Figure 5.21. High-pressure cube-optical apparatus. Section through anvil (after Owen[21])

Pressure is applied by a double cone arrangement in this case. The line of symmetry is a vertical body diagonal of the included cubic volume. In this way the anvils fit in two sets of three into two identical cones which mirror image about a horizontal axis (*Figure 5.20*). The cones are 15 cm outside diameter and 5 cm deep and are made from EN 26 steel hardened and tempered to R.C. 42.

108

Figure 5.22. Cubic anvil device (after Owen[21])

(To face p. 108)

The included cone semi-angle is 34° 15′ which is 1° less the theoretical value to reduce the effect of friction and to ensure that the line of loading is as near as possible the axis of the anvil (page 92).

In order to facilitate the assembly of the six anvils a 5 cm cube of Tufnol (laminated plastic) is bored out across each pair of opposite faces to the diameter of the anvils. By arranging the anvils in this guide block all six may be placed in position in the lower cone before the upper one is assembled (*Figure 5.22*). Two dowel pins are used to locate the cones.

As stated above this device has been built for use as an optical absorption cell, although it may be operated in a conventional way to do experiments similar to those in the tetrahedral apparatus. The optical path consists of a series of stepped holes through two opposite anvils and both cones as shown in *Figure 5.21*. The hole sizes in the cones are 1 cm and those in the anvils are telescoped down to a final 1 mm aperture. More than two anvils may be perforated and used to carry thermocouples into the high pressure region. The hole in the cone is larger than the hole in the anvil with which it is in direct contact to allow for relative movement under load. It is preferable to have three holes in each cone in order to create symmetrical straining conditions. The resulting aperture through the whole cell is about f. 8. The locating pins together with the Tufnol guide block ensure that the optical path is coaxial. The sliding friction between the backs of the anvils and the cones is reduced by inserting sheets of teflon as in the case of the tetrahedral apparatus. One of the most useful aspects of the device is the provision for six electrical probes by insulating the anvils from each other with sheets of mylar.

The pressure cell is normally a cube of pyrophyllite 7·5 mm side dimension which has been found to give maximum efficiency in respect of pressure generation for a given load. It is drilled out with a 2·5 mm hole across opposite faces (*Figure 5.23*). The materials used for the windows and for the pressure medium directly surrounding the sample depend obviously on the spectral range of interest and will be considered in greater detail below. The windows are polished to optical flatness to ensure good contact with the anvil faces. The pressure cell is assembled with the windows, pressure medium and sample inside the pyrophyllite cube. Again, as in the case of the tetrahedral device, buttons of teflon 1 mm thick are lightly attached to the anvil edges to correctly locate the oversized pyrophyllite cube as load is applied. It has been found that the Tufnol unit maintains the anvil alignment quite easily and no further guide system is necessary in the cones. As previously stated a third anvil is perforated

109

to take a thermocouple in the method of Boyd and England[22]. In this a specially formed retainer of stainless steel is used to prevent the thermocoax thermocouple wire from being ejected out of the high pressure cell. If a heater is required it may be made in the usual way with a graphite or metal sleeve of about 0·25 mm wall thickness. The maximum temperatures obtainable are less than those in the tetrahedral device because of the small length of the heater. As a rough guide approximately 1 watt dissipated generates 1° in the high pressure cell.

Up to 50 kb pressure the only sign of anvil deterioration is a slight cracking on the faces after 10 or 20 runs, and they remain usable for many more runs.

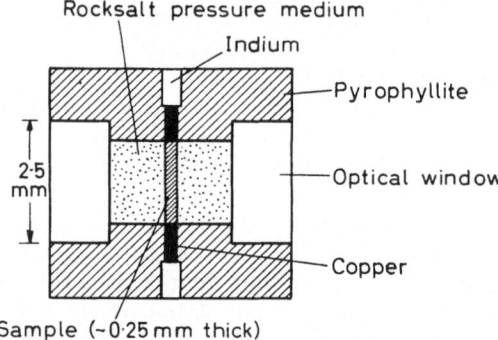

Figure 5.23. Pyrophyllite cube assembly for simultaneous optical and electrical resistance measurements

Calibration

The apparatus is calibrated in the usual way with the bismuth I–II, thallium and barium transitions which occur at approximately 20,000 kgf, 31,000 kgf and 60,000 kgf load respectively using an 7·5 mm pyrophyllite cube. The cell assembly for the calibration is straightforward with the appropriate wires inside a silver chloride sleeve in face to face configuration. Usually a reproducibility of about 3 per cent can be obtained.

In the case of pressures below 20 kb it is possible to calibrate by immersing a small manganin wire resistance gauge inside a capsule containing a liquid, for example mineral oil or pentane.

Optical Absorption Experiments

The window materials are defined by the spectral region which is to be investigated. *Figure 5.24* illustrates the relative transmission

110

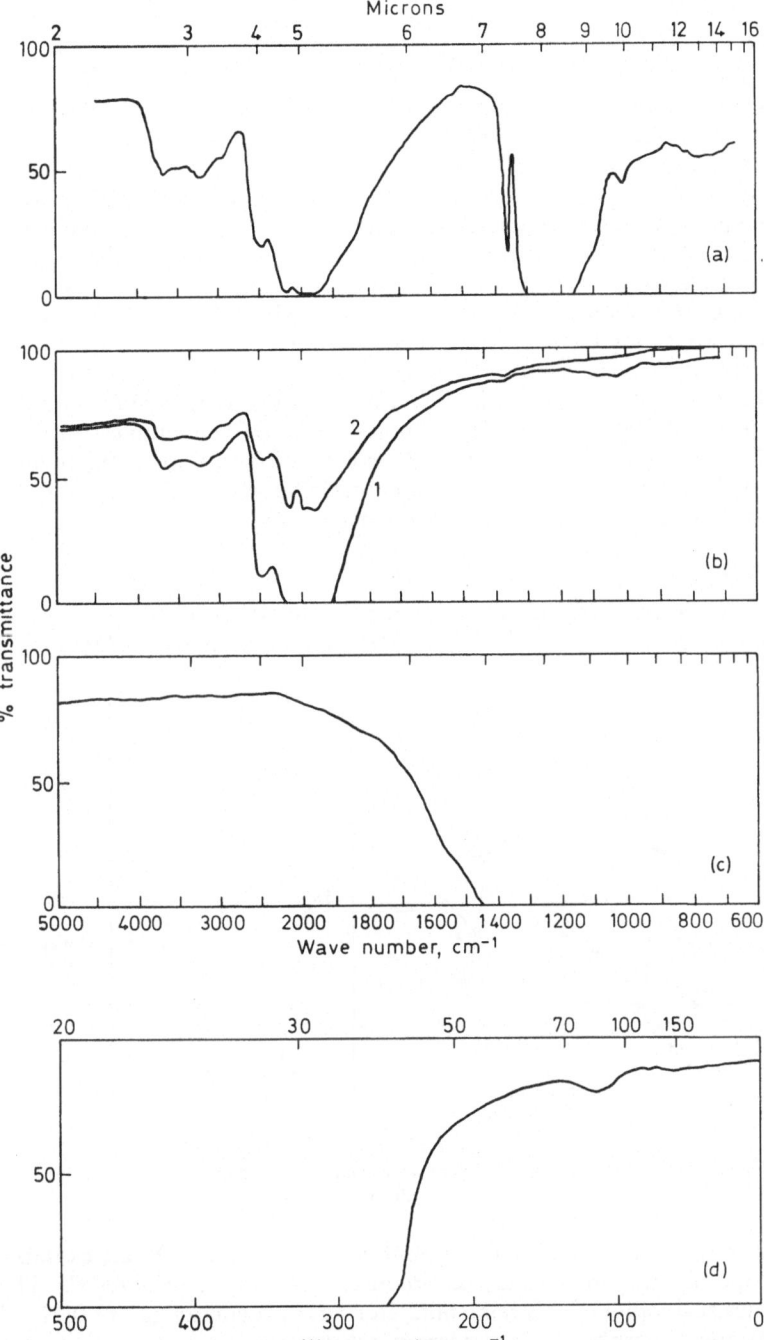

Figure 5.24. Relative optical transmissions. (a), type I diamond 2·5 mm thick; (b), type II diamond 2·5 mm (1) and 1 mm thick (2); (c), pure sapphire 1 mm thick; (d), single crystal quartz 1 mm thick.

of the three most commonly used materials, diamond, sapphire and single crystal quartz, which are chosen because of their suitable mechanical properties. The relative strengths are in the order diamond, sapphire and the comparatively weak crystal quartz. Type 1 or type 2 diamond windows may be obtained commercially in the form of small discs with two faces polished to a high degree of parallelism. Sizes about 2·5 mm diameter and 1·5 mm thick have been found to withstand up to 50 kb with a slight degree of cracking and have been used for between 10 and 20 runs before transparency is reduced to below tolerance level. These diamonds have usually a brownish colouration which is not important for infra-red experiments. The window seals are the simplest form of Poulter packing (see Chapter 3) and it is found that at the highest loads the anvils are slightly indented. The seal is always successful and no extrusion has been experienced.

Near infra-red absorption

An optical arrangement for the examination of energy gaps in semiconductors in the near infra-red region is shown in *Figure 5.25*

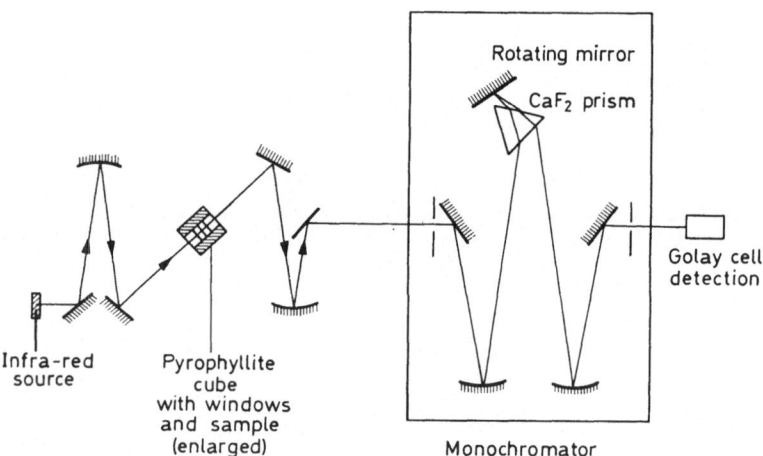

Figure 5.25. Arrangement for near infra-red spectroscopy using a cubic anvil device

and the pyrophyllite cube assembly in *Figure 5.23*. As an example experiments on indium antimonide will be considered[23]. The purest samples of this semiconductor have an energy gap of 0·17 eV at one atmosphere which corresponds to an optical absorption edge

112

frequency of about 7 microns. This wavelength is longer than those which can be investigated conveniently in the Drickamer optical cell. Indium antimonide crystals are prepared by polishing to optical flatness with diamond powder thin pieces about 0·25 mm thick. These are cut into discs approximately 2·5 mm diameter and pieces of single crystal rocksalt fit between either side of the indium antimonide and the diamond windows. (Rocksalt (sodium chloride) is transparent in this region.)

The absorption spectrum using a globar infra-red source is recorded from the output of the monochrometer at each load setting. A load maintainer for the press is advantageous here since the spectrum at a given pressure can take up to an hour to record.

Indium antimonide undergoes a phase change to the metallic state at 25 kb and this may be monitored using the electrical connection shown in *Figure 5.23*. A combination of indium which is soft and copper is used for the contact since direct contact of a hard material tends to intensify the pressure at the sample. It is found that provided the phase change point has not been reached the single crystal of indium antimonide remains intact as inferred from the reproducibility of the light transmitted after pressure cycling. This to some extent indicates a considerable degree of hydrostaticity at the centre of the cube.

Far infra-red absorption

The requirements of high-pressure containment and far infra-red radiation transmission are mutually contradictory. The first requires small apertures for maximum strength and the second large ones for maximum throughput of energy from the comparatively weak sources of far infra-red radiation. Experiments in this spectral region are extremely useful because absorption generally arises from lattice rather than electronic and molecular properties, the latter predominating in higher frequency regions. Lattice properties are most easily altered by pressure (and temperature) and far infra-red transmission experiments provide information complementary to that derived from x-ray diffraction and which is not obtainable by other means[24].

In order to overcome the built-in disadvantages stated above, the technique of Fourier transform spectroscopy is used. A description of the method is given by Gebbie[25]. Generally speaking spectrometers may be divided into two classes, monochromators in which information is recorded from each resolution element successively, and multiplex types in which all the information is recorded simultaneously. When the signal from a detector is noise limited, which is invariably the case in high pressure far infra-red experiments, the second of these spectrometers is extremely advantageous. A detector signal integrates linearly with time while noise integrates as the square root. Hence, if there is a total time t for observation of m resolution elements, in the monochromator method the time for each element is t/m and the signal to noise ratio is proportional to $\sqrt{(t/m)}$ and in the multiplex method the time taken for each element is t and the signal to noise is proportional to $\sqrt{t}$. Therefore there is a gain, known as the Fellgett gain, in signal to noise equal to $\sqrt{m}$ for the multiplex spectrometers. Typically an absorption experiment will be over the range 10 cm^{-1} to 200 cm^{-1} with 0·5 cm^{-1} resolution, giving a gain in signal to noise ratio of $\sqrt{380}$. There are many other factors to be taken into account and the above calculation is meant only as an illustration.

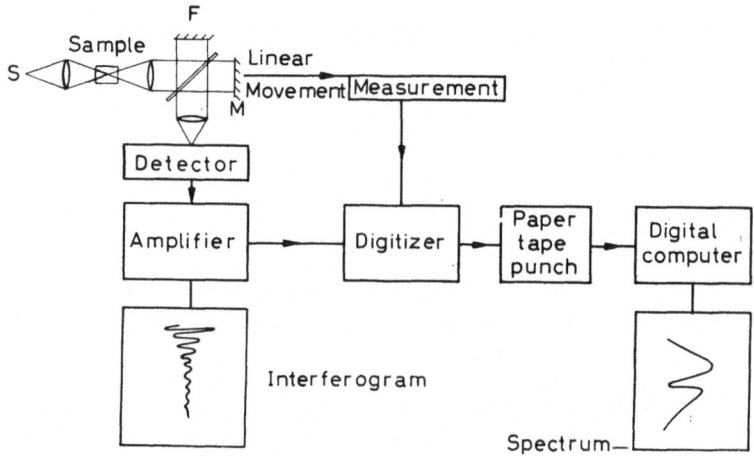

Figure 5.26. Arrangement for infra-red spectroscopy using a cubic anvil device

The multiplex spectrometer used with the cubic anvil apparatus is a Michelson interferometer as shown in *Figure 5.26*. The interference pattern or interferogram $I(\Delta)$ for monochromatic energy of frequency k has the form

$$I(\Delta) = G_0(1 + \cos 2\pi k\Delta)$$

where Δ is the path difference.

For a wide band source of spectral distribution $G(k)$.

$$I(\Delta) = \int_0^\infty G(k)[1 + \cos 2\pi k\Delta]dk \quad \text{(ignoring constant factors)}$$

114

From which

$$G(k) = \int_0^\infty [I(\Delta) - \tfrac{1}{2}I(0)] \cos 2\pi k \Delta \mathrm{d}\Delta$$

$I(0)$ is the intensity at zero path difference. $I(\Delta)$ is recorded as a function of Δ and $G(k)$ may be computed from the above equation[25]. The experimental arrangement shown in *Figure 5.26* has been used in the region 200 cm^{-1} to 10 cm^{-1} with single crystal quartz windows. In order to maximize the radiation throughput the holes in the anvil faces are increased to 2·5 mm diameter and although diamond would be the most suitable in view of its strength properties the cost per window is prohibitive. A cell assembly for studying infra-red spectra of liquid and crystalline organic materials is shown in *Figure 5.27*. The pressure limit with the larger holes is of the order of 40 kb and at this level the quartz windows become very crazed although they still retain liquids under load.

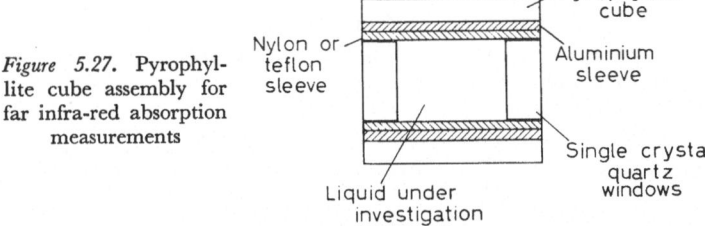

Figure 5.27. Pyrophyllite cube assembly for far infra-red absorption measurements

The cube anvil apparatus has several advantages over the Drickamer optical cell. It may be used at longer wavelengths; it is much easier to assemble and dissemble and to make simultaneous resistance measurements; and heating is a more straightforward process. The main disadvantage is the comparatively low upper pressure limit of 50 kb.

OTHER TYPES OF MULTI-ANVIL APPARATUS

Large Devices of Cubic Design

These have been constructed by Barogenics Incorporated[26] for commercial purposes and have anvil faces up to several cm in size. The multi-ram system is used with 1inch pins and hinges in a similar manner to that described for the tetrahedron. The pressure efficiency and uniformity has been studied by Brayman and Zeitlin[26] and it is claimed that the volume of uniform pressure is much greater

115

than that found with tetrahedral configuration. The apparatus is on an extremely large scale and although it has considerable use for commercial synthesis, it has little application for normal research laboratories. Another large scale cubical device is in operation in the High Pressure Research Institute in Moscow using individual ram loading[27].

Von Platen's Cube Device[28]

This is illustrated in *Figure 5.28*. It has been used for the manufacture of synthetic diamonds. Six anvils are driven by rams which form part of a sphere. They are surrounded by a copper shell in which pressures of 6 kb can be contained. The load multiplication factor is about 20 and so pressures up to 120 kb are generated. The copper shell is supported by a wire wound sheath which is a method used extensively by the Swedish A.S.A.E. Co. The high pressure volume is about 400 cm³ and in the form of a copper cube with steel bridges to prevent extrusion. A considerable part of the load is taken by the gaskets and the pressures inside the cell are probably much less than 120 kilobars. Temperatures of 4,000°C may be obtained by means of a heater circuit through an insulating sleeve in one anvil.

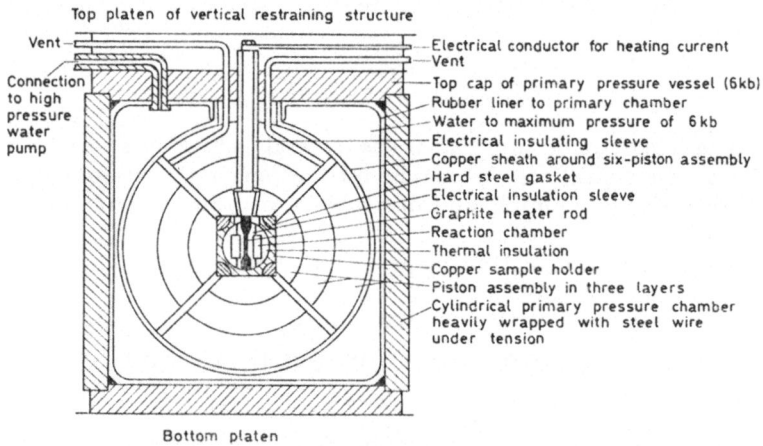

Figure 5.28. Von Platen's cube device

Japanese Cubic Device[29]

This is very similar to the N.B.S. and N.P.L. construction in so far as load is applied to six anvils by means of a uniaxial force and wedge reaction. *Figure 5.29* illustrates the essential difference. Four anvils are situated in guide blocks in the horizontal plane and the back faces are in the form of a 45° truncated wedge. The upper and

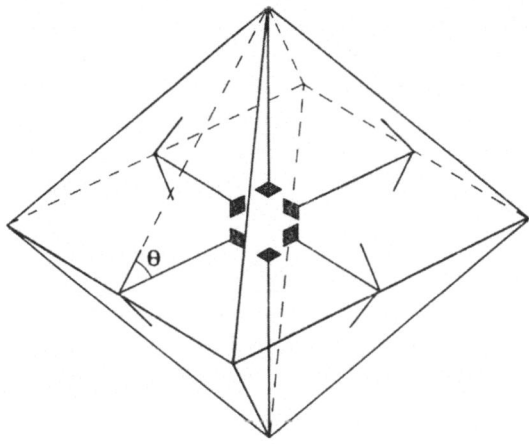

Figure 5.29. Cubical anvil device after Osugi *et al.*[29]

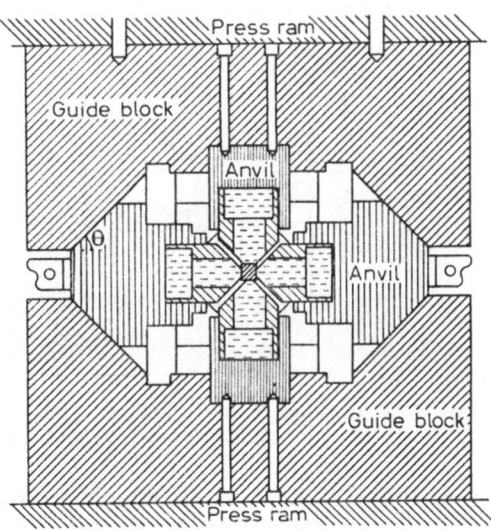

lower anvils are fixed in guide blocks whose inner surfaces are tetrahedral bi-pyramids. Load is applied vertically and is converted to six equivalent components acting on the faces of the pyrophyllite cube. The advances of the anvils are therefore automatically

117

equalized with increasing pressure as in the wedge cone technique. The anvils are of tungsten carbide and have square faces from 1·5 up to 3 cm. Calibrations in the usual manner have shown that the apparatus is capable of reaching the upper bismuth transition and higher and using the usual techniques temperatures of 2,000°C may be obtained. The insertion of a fine copper wire mesh inside the pyrophyllite cube in a number of geometrical positions has shown that the pressure is uniform throughout a large proportion of the volume. To minimize friction and to provide electrical insulation the usual sheets of teflon and melinex are inserted between each anvil and the two guide blocks.

Philips High Pressure Device[30]

This apparatus is a gasket compression type with cylindrical symmetry. Load is applied through a high-pressure water tyre constrained by an outer steel shell as in the case of Von Platen's

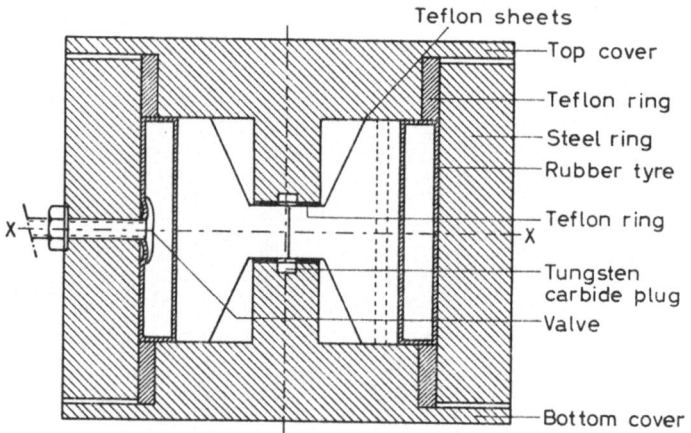

Figure 5.30. Drawing of the cylindrical high-pressure apparatus (after Witteman and Werkman[30])

cube device. The pressures obtainable are up to 100 kb in a comparatively large volume. As the apparatus does not require a press of very high capacity and is comparatively simple to construct it is worth describing in some detail. The construction is illustrated in *Figures 5.30* and *5.31*. Pressure is generated in the volume enclosed by three cylindrical tungsten carbide rods. An oversized pyrophyllite or similar substance sample cell is used in the usual way to form compressible gaskets. The tungsten carbide cylinders are supported over an angle of 180° by hardened steel segments into which they are press-fitted. The high pressures are generated over a sector subtending an angle of 60° to the centre. The three segments are held in position by three sliding pieces whose backs like those of the segments have cylindrical curvature to fit the water tyre. The latter is capable of withstanding pressures up to 2·5 kb generated by a small hand pump. At the top and bottom of the apparatus two plates with tungsten carbide inserts apply end loading to the cylinders and provide a constraining surface for the top of the water tyre. A teflon sheet on top of the cylinders ensures a minimum of friction resistance as the segments move inwards under load. As will be seen from the section of the apparatus in *Figure 5.30* there is a pressure multiplication effect from both the differences in area and from the reduction in height in the central region.

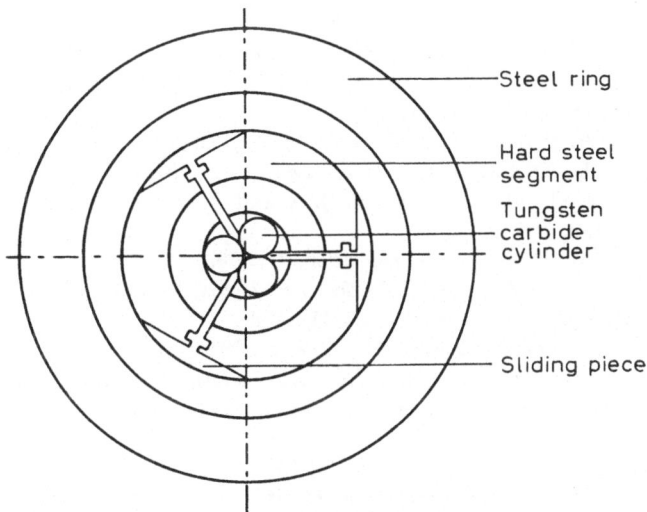

Steel ring

Hard steel segment

Tungsten carbide cylinder

Sliding piece

Figure 5.31. Section through XX of the apparatus in *Figure 5.30*
(after Witteman and Werkman[30])

The operation of the device under load has been analysed in detail[30] and it has been shown that the pressure is uniform along the axis of the cylinders and that the compressibility condition of the high-pressure volume is independent of the axial length. To take an example, an apparatus with carbide cylinders 1·5 cm diameter and height 2·8 cm has a pressure multiplying factor of 42. The reader is referred to the original publication for more details.

REFERENCES

[1] Hall, H. T. *Rev. sci. Instrum.* 1958, **29,** 267.

[2] Houck, J. C. and Hutton, U. O. *High Pressure Measurement,* p. 221. Eds. A. A. Giardini and E. C. Lloyd. Butterworths, London, 1963.

[3] Vereshchagin, L. F. *Progress in Very High Pressure Research.* Eds. F. P. Bundy, W. R. Hibbard and H. M. Strong. Wiley, New York, 1960.

[4] Lees, J. *Advances in High Pressure Research,* vol. 1, p. 1, Ed. R. S. Bradley. Academic Press, New York, 1966.

[5] Lloyd, E. C., Hutton, U. O. and Johnson, D. P. *J. Res. Nat. Bur. Stand.* 1959, **63c,** 59.

[6] Lees, J. *Nature, Lond.* 1964, **203,** 965.

[7] King, J. H. *J. sci. Instrum.* 1964, **41,** 102.

[8] King, J. H. *J. sci. Instrum.* 1965, **42,** 374.

[9] Lees, J. and Williamson, B. J. H. *Nature, Lond.* 1965, **208,** 278.

[10] Jayaraman, A., Klement, W., Newton, R. C. and Kennedy, G. C. *J. Phys. Chem. Solids,* 1963, **24,** 7.

[11] Butozov, V. P. *Sov. Phys. Crystallogr.* 1957, **2,** 533.

[12] Pugh, H. Ll. D. and Lees, J. *Nature, Lond.* 1961, **191,** 865.

[13] Butcher, E. G., Alsop, M. Weston, J. A. and Gebbie, H. A. *Nature, Lond.* 1963, **199,** 756.

[14] Claussen, W. F. *Rev. sci. Instrum.* 1960, **31,** 878.

[15] Blum, F. A. and Deaton, B. C. *Phys. Rev. Letters,* 1964, **12,** 697.

[16] Hall, H. T. *Rev. sci. Instrum.* 1962, **33,** 1278.

[17] Zeitlin, A. *Scientific American,* 1965, **212** (5), 38.

[18] Barnett, J. D., and Hall H. T. *Rev. sci. Instrum.* 1964, **35,** (2), 175.

[19] Jeffery, R. N., Barnett, J. D., Vanfleet, H. B. and Hall, H. T. *J. appl. Phys.* 1966, **37,** 3172.

[20] Decker, D. L. *J. appl. Phys.* 1965, **36,** 157.

[21] Owen, N. B. *J. sci. Instrum.* 1966, **43,** 765.

[22] Boyd, F. R. and England, J. L. *J. Geophys. Res.* 1960, **65,** 741.

[23] Bradley, C. C. and Gebbie, H. A. *Phys. Letters,* 1965, **16,** 109.

[24] Bradley, C. C., Gebbie, H. A., Gilby, A. C., Kechin, V. V. and King, J. H. *Nature, Lond.* 1966, **211,** 839.

[25] Gebbie, H. A. *Advances in Quantum Electronics.* Ed. J. R. Singer. Columbia University Press, New York, 1961.

[26] Zeitlin, A. *A.S.M.E. 60-WA-333,* 1961.

REFERENCES

[27] Vereshchagin, L. F. *Progress in Very High Pressure Research*, p. 290. Eds. F. P. Bundy, W. R. Hibberd and H. M. Strong. Wiley, New York, 1961.

[28] Bundy, F. P. *Modern Very High Pressure Techniques*, p. 9. Ed. R. H. Wentorf. Butterworths, London, 1962.

[29] Osugi, J., Shimizu, K., Inoue, K. and Yasunami, *Rev. Phys. Chem. Japan*, 1964, **34**, (1), 1.

[30] Witteman, W. J. and Werkman, T. *Philips Res. Reports*, 1963, **18**, 447.

121

6

PISTON AND CYLINDER APPARATUS FOR PRESSURES UP TO 100 KILOBARS

In this chapter a description is given of a number of devices with the same design specifications as the cubic and tetrahedral anvil apparatus, that is, pressures and temperatures up to 100 kb and 2,000°C in volumes of the order of several cm³. These include conventional piston and cylinder apparatus using solid pressure-transmitting media but in all other respects similar in design to the hydrostatic apparatus described in Chapter 3 and the 'belt' and 'girdle' devices which are not strictly speaking piston and cylinder types, but since the pressure regions usually have cylindrical symmetry they have been included in this chapter. The generation of pressure is different in the two cases. In the first it is straightforward force over area in a rigid system, in the second the compressible gasket method is used. The apparatus is simpler than the cubic and tetrahedral anvil types in many respects but the construction of the sample cells and gaskets is more complicated.

In extending the pressure range of piston and cylinder apparatus to the region of 50 kb and above both the cylinder and piston require additional support since the limiting strength of an unsupported tungsten carbide piston is about 50 kb and of a cylinder not much above 25 kb. In one of the first very high-pressure apparatus, Bridgman[1] enclosed a small piston and cylinder assembly capable of withstanding 30 kb inside a larger pressure cylinder. By subjecting the smaller unit to 30 kb hydrostatic pressures up to 100 kb were generated. The multiplication is greater than two because of the enhancement of the strength properties of steels under pressure. As will be seen from *Figure 6.1*, although this apparatus in principle is very useful, in practice the complicated assembly and difficulty in making measurements prevent it from becoming widely used.

In a number of devices described below the pistons are provided with independent support generated in a second-stage compression chamber. In this way a piston may be supported over most of its length outside the main cylinder.

A major modification is to change the standard piston and cylinder profiles and to use the principle of massive support for both. By combining with support from a compressible gasket extremely high pressures are generated. The 'belt' and 'girdle' apparatus evolved through this method.

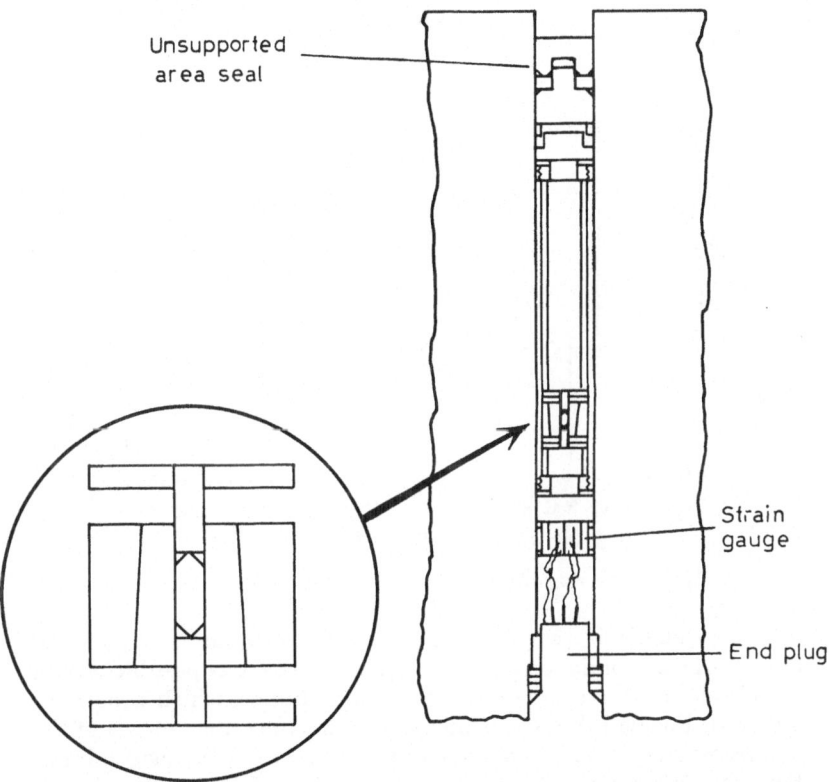

Unsupported area seal

Strain gauge

End plug

Figure 6.1. Left: The miniature pressure vessel for producing 100 kb. This is mounted inside the 30 kb apparatus. Right: Details of the mounting of the 100 kb vessel within the 30 kb vessel (after Bridgman[1])

BINDING RING SUPPORT FOR CYLINDERS

In order to obtain maximum support for a cylinder under pressure a series of supporting rings are either shrunk or press-fitted onto the cylinder. The design of a set of rings follows well-developed lines and several versions have been published[2,3]. A mathematical analysis of a required stress system for a given size and materials may be made but the results are rather indeterminate and the best procedure is to make use of the experience of a number of laboratories. The version described here has been used at the National Physical Laboratory and is illustrated in *Figure 6.2*.

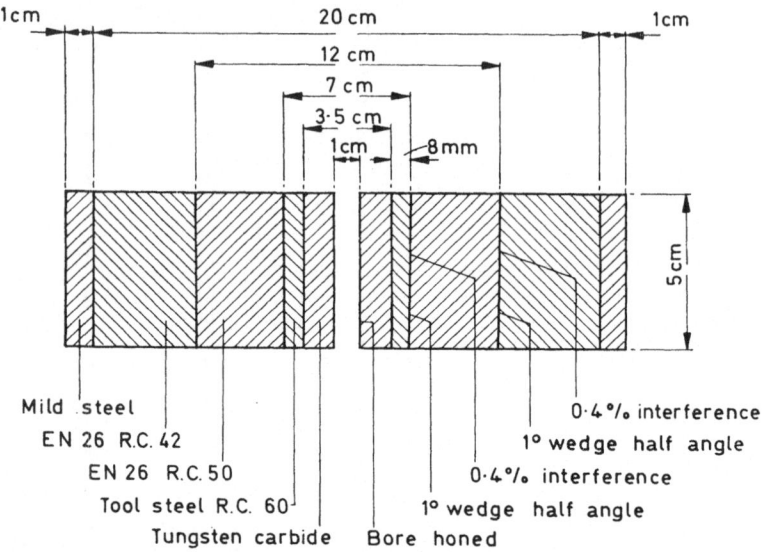

Figure 6.2. Binding ring assembly for high pressure cylinder (not to scale)

The principle is that of prestressing the cylinder such that its inner surface is under as high a compressive stress as possible within the elastic limit and creating a stress distribution which progresses outwards through the supporting rings until the outer surface of the outermost ring is under tensile stress, again within the elastic limit. For most purposes no more than two supporting rings are used. The stresses are usually set up in hardened steel rings by forcing one inside another with a wedge-shaped profile such that the outer diameter of the inner ring is slightly larger than the inner diameter of the outer ring. The difference between the diameters of the two surfaces in contact is known as the interference and is usually expressed as a percentage of the average diameter. Since there is considerable stored energy in such a stress system an outer safety ring of mild steel is always press-fitted around the second binding ring. An alternative method is to heat the rings to a temperature at which they can be pushed together easily and the stress is set up on cooling down. However, the temperature limit is set by the properties of the steels (see Chapter 2) and in normal cases the

124

interference obtainable by this means is much less than by mechanical press-fitting. The example shown in *Figure 6.2* is for support of a tungsten carbide cylinder 1 cm diameter bore, 5 cm long, and 3·5 cm o.d. The supporting rings are both constructed of EN 26 steel, hardened and tempered to R.C. 42 and 50 respectively. The outer ring, which is approximately 20 cm o.d., is pressed first into the mild steel ring and then the inner ring with 0·4 per cent greater diameter is forced into the outer ring using molybdenum disulphide as lubricant. The wedge half-angle is approximately 1° and is such that it is half the angle of friction between the two surfaces with molybdenum disulphide. If a larger angle is used there is a chance that the ring will spring out. (Larger angles have been used but presumably with greater friction between the surfaces in question.) The force needed to press the rings together with 0·4 per cent interference may be up to 100,000 kgf for the size of the apparatus shown. The exact value for the interference is based purely on experience within the obvious limit that there is no fracture caused in pushing the rings together. However there is a considerable degree of plastic deformation of the steel. The diameters of the supporting rings are chosen so that the outer to inner diameter ratios are approximately the same since this helps to produce a more uniform stress variation across the whole assembly. A shim consisting of four segments of tool steel (R.C. 60) is placed between the surfaces of the inner ring and the tungsten carbide cylinder. The final stage is the simultaneous press-fitting of the shim and cylinder into the inner ring. There is no interference fit between the shim and carbide cylinder but that between the inner supporting ring and the shim is approximately 0·4 per cent on a 1° wedge. The shim provides an easy method of removing a broken carbide cylinder from the assembly without having to take any of the binding rings apart. The bore of the cylinder is not finally ground and lapped until after this final stage since any plastic deformation due to the compressive forces may then be removed.

PISTON AND CYLINDER APPARATUS

Boyd and England Single-Stage Apparatus[4]

This is illustrated in *Figure 6.3* and pressures up to 50 kb may be generated in a solid pressure medium. The pressure vessel is a tungsten carbide cylinder supported by steel binding rings in a similar manner to those shown in *Figure 6.2*. Additional support to prevent lateral fracture is provided by end loading the cylinder in a press with a capacity of several 100,000 kgf. (Alternatively, plates

bolted together can be used to provide end loading, see *Figure 6.5.*)
The piston, also of tungsten carbide, is ground and lapped such
that it fits the cylinder bore with less than 0·015 mm clearance.
It is supported over part of its length by a press-fitted steel ring.

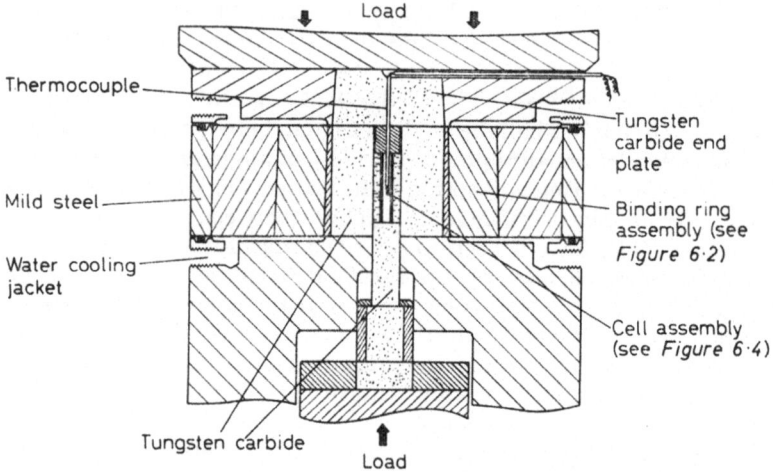

Figure 6.3. Single-stage piston cylinder (after Boyd and England[4])

A second press capable of up to 100,000 kgf is required for generating the pressure. Provision for water or liquid nitrogen cooling is
incorporated as shown.

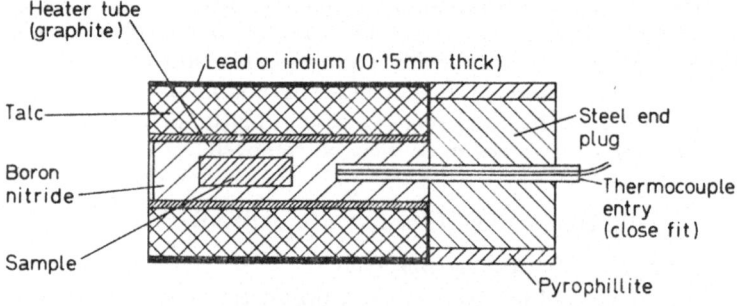

Figure 6.4. General cell assembly for piston and cylinder apparatus

A frequently used cell assembly is shown in *Figure 6.4.* Talc is
sometimes preferred to pyrophyllite or boron nitride as the pressure
medium since its internal friction is less than pyrophyllite and its

126

thermal conductivity is less than that of boron nitride. The end plug is a stainless steel cylinder with a small hole through which ceramic tubing carries thermocouple or resistance leads. Phosphor bronze or beryllium–copper delta-rings are usually used between the piston and the sample cell to minimize extrusion past the piston. The plug is insulated from the cylinder by a thin pyrophyllite sleeve or other appropriate material. Mylar sheets are placed in appropriate positions to provide a heating circuit via the piston and the end plug. In order to reduce friction the talc cylinder may be enclosed in a thin sheet of lead or indium (0·15 mm thick) although care must be taken to avoid electrical contact with the steel end plug. Anti-extrusion rings of steel or phosphor bronze prevent the soft metal extruding past the piston. Up to a hundred runs over the temperature and pressure range 1,500°C and 50 kb have been obtained with a single carbide cylinder. In the apparatus shown in *Figure 6.4* a heater efficiency of approximately 1° per watt with a graphite sleeve 0·4 mm wall thickness is usual. Calibrations indicate that at room temperature friction accounts for about 13 per cent of the load, most of this being inside the talc. At higher temperatures this is considerably reduced.

The apparatus can be used in quite simple form up to 25 kb, that is without end loading the tungsten carbide cylinder and using straightforward simply constructed pistons.

Boyd and England Two-Stage Apparatus[5]

The above apparatus is limited to 50 kb or less because of fracture of the pistons which are supported only over a small fraction of their lengths. The pressure range may be extended to approximately 100 kb by the introduction of a second stage to provide support for the piston. This is illustrated in *Figure 6.5*. Most of the apparatus is similar to that already described but the high-pressure piston has a separate tapered portion which advances under load through a second pressure vessel in which is packed a highly compressible substance such as potassium bromide. The dimensions depend on the cell assembly and have to be calculated for a given case. End load support is provided for the double cylinders as before. The thrust on the large piston is divided between that on the packing material in the supporting stage and on the piston in the high-pressure cylinder in the ratio of approximately 1 to 4 for an apparatus of the size indicated in *Figure 6.5*. Friction is reduced in the supporting stage by liners of indium or lead and anti-extrusion rings are used appropriately. A suitable calibration technique is to determine a compression curve without a high pressure sample-assembly and

127

then repeat with the latter present. Using the values for the displacement of the piston in each case the thrust on the packing material can be calculated and therefore subtracted from the total load to give the load on the high-pressure assembly. After allowing for friction the pressure can then be determined. The limit is reached in potassium bromide by a transition at 18 kb but this may not be inconveniently low since the high-pressure pistons begin to deform plastically in the 75 kb region. Different packing materials such as bismuth and pyrophyllite have been used with apparent success. The failure of the apparatus arises usually from radial and lateral cracking of the tungsten carbide cylinder.

Other two-stage piston and cylinder apparatus has been built by Giardini et al.[6] and Bradley et al.[7]

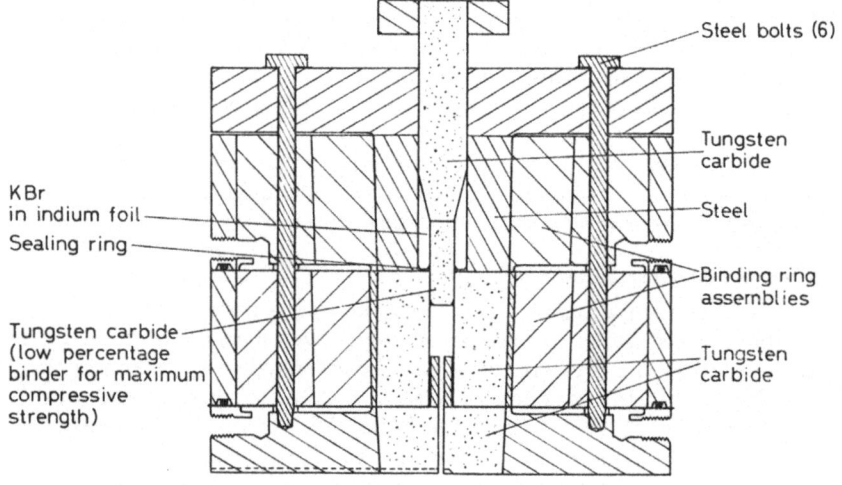

Figure 6.5. Double stage piston and cylinder (after Boyd and England[4])

THE 'BELT' HIGH-PRESSURE APPARATUS

The 'belt' apparatus may be described crudely as a cross between a piston and cylinder and Bridgman anvils. It combines massive support for the pistons with a central containing die for the sample. The pressure generation is through the compression of a gasket material as in the case of the tetrahedral and cubical devices. The 'belt' is probably the most useful apparatus available for ultra high-pressure/high-temperature experiments. In it pressures greater than 100 kb and temperatures up to 2,000°C can be held simultaneously for many hours in sample volumes of the order of several cm³. It was developed at the General Electric Company by Bundy, Hall and Strong and others for use in diamond synthesis[8]. The restrictions imposed on the publication of the details of its construction because of the strategic importance of diamond synthesis led to the development of a number of closely related devices.

The 'belt' evolved from Bridgman anvils (Chapter 4) by first incorporating a well in one of the anvils and compressing a pyrophyllite assembly. However, the pressure limit is quickly reached when the pyrophyllite extrudes to its limiting thickness and halts further compression inside the well. The major advance in design was made with the realization that with a conical profile the compression of the gasket is reduced by a factor of the sine of the half-angle of the cone for a given vertical compression stroke[8]. Thus considerable compression of the sample may be obtained before the limiting thickness of the gasket is reached. The final step is to double-end the apparatus about a horizontal axis producing a device consisting of two anvils or pistons compressing a sample restricted in a central die. After further sophistication the final version is the 'belt' apparatus.

The detailed account which follows is taken from U.S. Patent No. 2,941,248 taken out by G.E.C. Since the apparatus is completely symmetrical about a horizontal axis through the centre it will be sufficient to describe just a single piston and the upper part of the die.

The piston is constructed from tungsten carbide and has a high-pressure bearing face approximately 0·8 cm diameter. The profile of the piston between this surface and the load-bearing surface which is approximately 3·5 cm in diameter is as shown in *Figure 6.6*. The initial conical portion with a half-angle of 30° to the vertical extends for about 0·5 cm and then there is a smooth curve to the upper surface. The distance between the lower and upper surfaces is about 1 cm and should be kept to a minimum for maximum strength. The piston is supported in three ways on its pressure bearing surface, as follows.

(a) By massive support of the conical section.

(b) By press-fitted supporting rings of hardened steel (for example R.C. 50 EN 26). As will be seen in *Figure 6.6* two hardened steel supporting rings are used with outer diameters about 10 cm and 14 cm respectively. The tungsten carbide insert has a 1·5° taper and a 0·7 per cent interference fit with the first steel ring. The interference fit between the two steel rings is about the same. A soft

129

steel outer ring is used as usual for safety purposes. The lower face of the ring assembly has a 7° taper with the horizontal, so that a smooth fit with the curved surface of the carbide insert is obtained. The total diameter is approximately 15 cm.

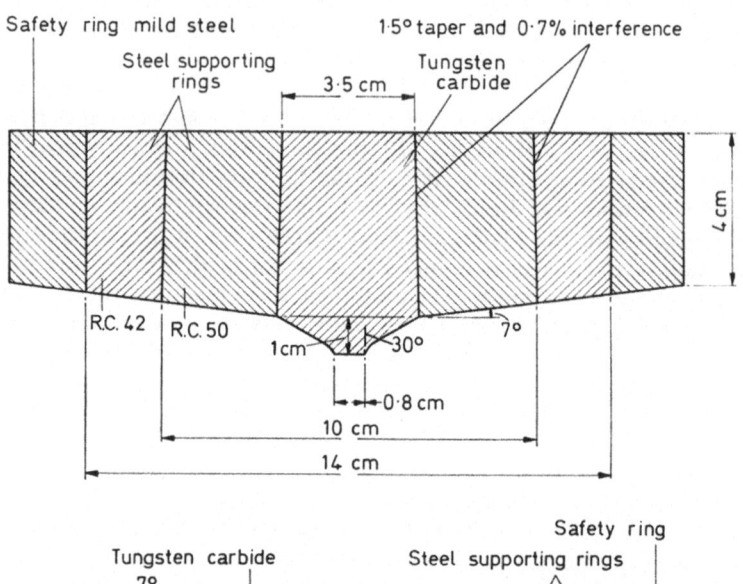

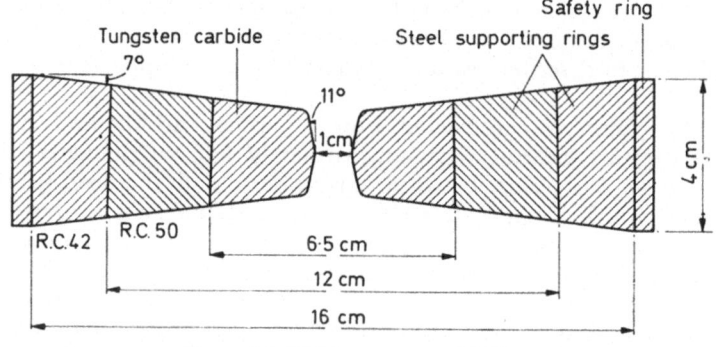

Figure 6.6. 'Belt' apparatus (Hall[8])

(*c*) By resolution of the vertical force through compression of a gasket between the curved surfaces of the piston and die.

The die is constructed of tungsten carbide and has an internal profile such that the tapered piston can move freely into it. The inner component is put into compressive stress by means of supporting rings in the usual manner. The die is approximately 6·5 cm in diameter and the hardened steel rings are 12 cm and 16 cm in

130

diameter respectively. The tapers and interference fit are similar
to those for the pistons although a slightly larger value for the interference between the first ring and the tungsten carbide insert is
used. The height of the die increases radially a by 7° taper which
helps in the support. The profile of the inner surface of the die is
as shown. The bore is approximately 1 cm at the centre and is
flared out at an angle of 11° to the vertical for the first 0·6 cm and
then there is a smooth fit to the 7° taper on the upper part of the
surface. When there is a compression of the gasket between the
two curved surfaces the hoop stress set up in the curved surface of
the die by pressure in the central region is counterbalanced by the
compressive force in the gasket.

The gaskets are usually constructed separately for the upper and
lower halves of the apparatus. They consist of two conical pyrophyllite gaskets sandwiching a hollow steel cone (*Figure 6.7*). The

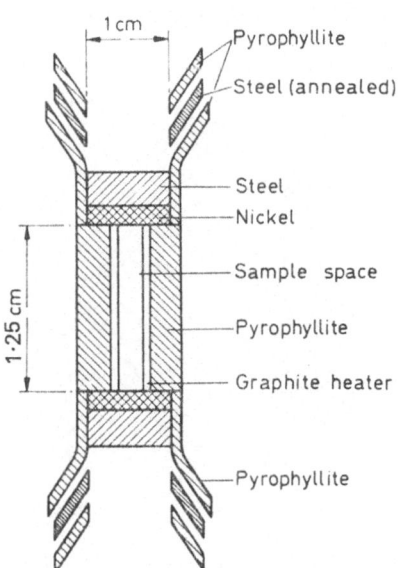

Figure 6.7. Sample cell and
gaskets for 'belt' apparatus
(after Hall[8]) (not to scale)

presence of the steel which should be annealed is important since
it prevents the thick pyrophyllite gaskets from breaking at low
pressures and helps to lower the compressibility of the combination.
The steel cone is about 0·25 mm wall thickness and the profiles of
the pyrophyllite gaskets are shaped so that they fit the available
space. A cylindrical region approximately 1 cm diameter and

1·25 cm high is left for the high pressure cell assembly which usually consists of samples and heaters inside a pyrophyllite cylinder. Further details of assemblies for particular purposes are given at the end of the chapter. Heating is usually by means of a metal or graphite sleeve and thermocouple leads may be inserted into the high-pressure regions through holes drilled in the thickest part of the gaskets. With the dimensions given here a total piston stroke of about 1 cm is possible. Temperature and pressure calibrations are carried out in the usual manner (Chapter 1).

Failure of the apparatus is usually by radial cracks in the central insert of the die but one of the major advantages of the 'belt' is that up to 100 kb the failure rate is very low compared with other apparatus. According to the patent when operating around 50 kb and 1,000°C the inner die shows fine cracks after 40 or 50 runs and another 35 are possible (runs are up to one hour duration). The pistons are not damaged after more than 100 runs. The use of hardened steel instead of tungsten carbide only slightly lessens the life.

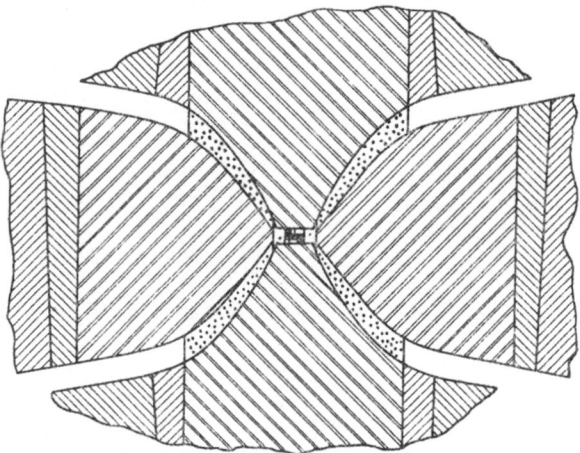

Figure 6.8. High compression 'belt' apparatus (after Bundy[10])

Modifications have been made to the supporting ring assembly and to the shape of the inner profile of the piston and die with some slight improvement in efficiency[9]. A high-compression version of the 'belt' has been described by Bundy[10] in which pressures of 200 kb can be generated. The main difference is in the profile as shown in *Figure 6.8.* The bore of the die is about 0·5 cm and the piston diameter 0·4 cm. The useful part of the pressure region is only 0·2 cm high.

'GIRDLE' HIGH-PRESSURE APPARATUS

This apparatus is based in form more on the Bridgman anvil configuration than the 'belt'. It is in fact quite similar in principle to the supported anvil Drickamer devices (Chapter 4). It has the advantage of requiring less elaborate machining than the 'belt'. In the first version which was built by Wilson[11] the apparatus takes the form of a central die with conical-shaped openings and central cylindrical profile about 1/3 the total height and two conical pistons of the same half-angle as the die. Support of the lower part of the piston is obtained from massive support due to the conical shape and reaction forces at the interface of the slanted faces. Compression of the samples is obtained by elastic distortion of the bore of the die with load on the pistons (*Figure 6.9*). The supporting rings around

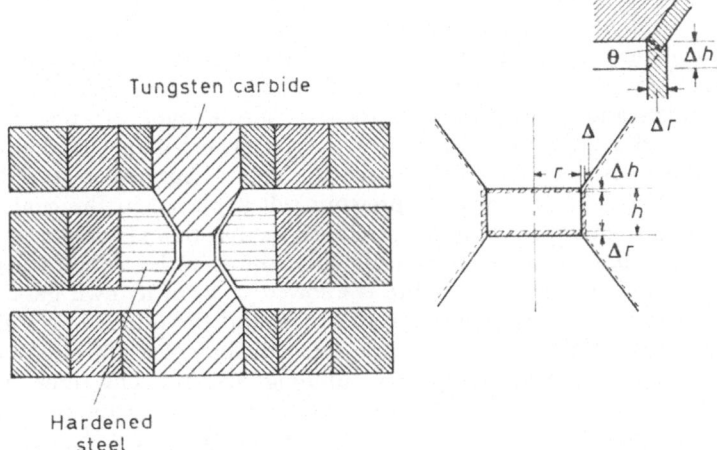

Tungsten carbide

Hardened steel

Figure 6.9. Girdle high pressure apparatus. The right-hand diagram shows the geometry operative in the girdle die design (after Wilson[11])

the hardened steel (R.C. 50) central die follow the usual pattern of interference fits as described above. Load is applied to the tungsten carbide pistons and they are driven into the conical recess in the die causing it to expand. If the decrease in volume due to the stroke of the pistons is greater than the increase in the expansion of the bore, then there is net compression of the central volume. Steel is

133

chosen as the material of the die because of its more suitable behaviour in this process. It will be seen from geometrical considerations in *Figure 6.9* that the threshold condition is that $r = h \tan \theta$ neglecting second-order terms. In other words the height h of the bore has to be less than $r/\tan \theta$ for the generation of pressure. The conditions used by Wilson in the original device were $r = 0.6$ cm, $\theta = 35°$ and h between 0.8 cm and 0.6 cm. The smaller value for h is advantageous since it results in a greater rate of pressure generation with piston stroke.

The net effect of the two conical surfaces coming together under load is one of mutual support as in the 'belt' but with the absence of any intermediate gasket material. Electrical insulation for heating or resistance measurement purposes is maintained by coating the surfaces of the upper piston with a thin layer of oxide. Wilson has shown that the friction between the surfaces adds appreciably to the support mechanism. The usual type of assembly for the central region is used, that is with metal sleeve heaters and pyrophyllite pressure medium. In order to avoid gasketting due to extrusion of the pyrophyllite, which would tend to produce weakening stresses, the cell may be pre-compacted and the excess material removed. With 1.25 cm bore dies pressures of 100 kb can be reached.

Daniels and Jones[12] have adapted a 'girdle' type of apparatus for use with a compression gasket. A pre-formed gasket (teflon or polythene) is shaped to fit between the conical surfaces of the die and piston. The pyrophyllite pressure cell is placed in the central region and an appropriate size is chosen so that there is only a very small amount of extrusion outside this region under compression. (It is found that if pyrophyllite is allowed to form its own gaskets the latter rapidly reach limiting thickness and the limit of pressure generation.) Using a central area 1.25 cm diameter and 1.25 cm high, plastic gaskets about 0.6 cm long and 0.2 cm thick are optimum. Daniels and Jones studied the efficiency of the apparatus as a function of cone semi-angle and found that between 25° and 50° there is little variation. The pressure/load efficiency is increased through the reduction of the overall compressibility by inserting small caps of hardened steel inside the high pressure cell. Although the presence of these caps increases the conduction of heat to the piston faces they do not appear to reduce their life. The Daniels and Jones apparatus is capable of reaching 100 kb and 3,000°C and approximately 65,000 kgf load is required to reach the first bismuth transition (25.4 kb) in a 1.25 cm diameter die. The principle of pressure generation in this apparatus is not the elastic distortion method of Wilsons' original 'girdle' and hence tungsten carbide may be used for the central die as well as for the pistons.

Young et al.[13] have combined the 'girdle' apparatus with the sandwich gaskets of the 'belt' (*Figure 6.10*). As described above these

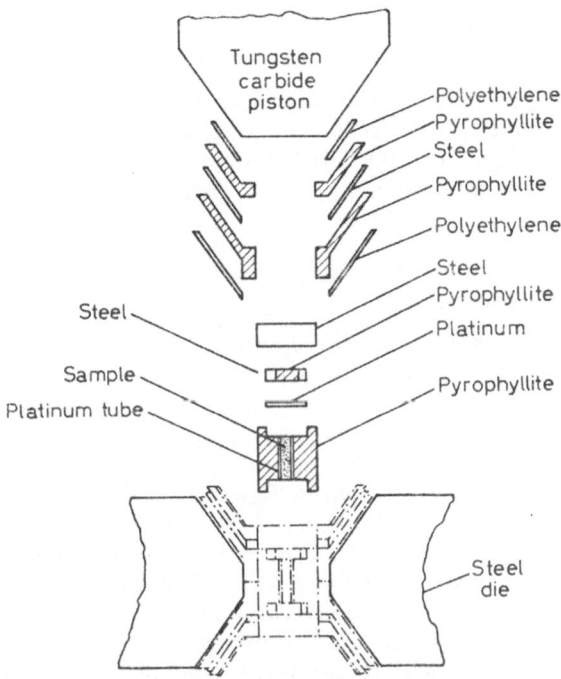

Figure 6.10. Schematic of specimen cell and gasket assembly in the revised girdle apparatus of Young et al.[13]

are sandwiches of soft steel cones and pyrophyllite. The pressure/load efficiency is almost doubled by the further insertion of 0·1 mm polythene sleeves at the interface of the pyrophyllite and metal surfaces of the die and pistons. The height of the bore is reduced to 0·6 cm and pressures up to 140 kb can be generated as indicated by resistance transitions in iron and barium (Chapter 1). Hardened steel caps are again used to decrease the overall compressibility of the cell. The mode of operation is a combination of elastic distortion of the die and compression of the gasket.

The limit of usefulness of this device over a wide range of temperature is set by the height of the sample volume since this automatically decides the heater size. However, quite high pressures should be possible with small sample heights in view of the capability of Drickamer's supported pistons of generating upwards of 400 kb (Chapter 4).

The apparatus fails with radial fracture of the hardened steel die and although exact figures are not available it is probable that many less runs to 100 kb are possible in the 'girdle' type than in the 'belt' type before failure of the die.

To summarize, the 'girdle' apparatus together with the straight-forward piston and cylinder has the advantage of simpler construction over the 'belt' but the latter gains in a higher pressure limit over the piston and cylinder and of longer component life than the 'girdle'.

CELL ASSEMBLIES FOR USE IN PISTON AND CYLINDERS, 'GIRDLE' AND 'BELT' APPARATUS

Resistance Measurements at High Temperatures

The cell assemblies which will be described here are of general form and may be used in most cases in any of the devices described in this chapter, after making suitable allowance for relative dimensions. Two assemblies for resistance measurements up to many hundreds of degrees are shown in *Figure 6.11*. In all cases the best medium for pressure transmission at high temperatures is pyrophyllite, pipestone, or possibly talc. Pyrophyllite is electrically insulating and has low thermal conductivity which is necessary to protect the metal parts of the apparatus from high temperatures. At high temperatures and pressures it transforms to a whitish powder consisting of kaynite and coesite but retains its rigidity for sample holding up to nearly 3,000°C. Silver chloride sleeves may be used to improve homogeneous pressure conditions around samples but it cannot be used above 400°C because of melting and increased reactivity. For higher temperatures the hexagonal form of boron nitride can be substituted but it is not usable at the highest pressure and temperature because of a transition to the cubic form and it also becomes more reactive. Heaters may be either metallic or graphite hollow cylinders about 0·3 mm wall thickness and contact is usually made to the pistons via a thin steel ring which surrounds a pyrophyllite plug acting as thermal insulation. The question of temperature and pressure calibration has been discussed in Chapter 1.

Phase Changes by Thermal Conductivity Measurements

Figure 6.12 illustrates the differential thermal conductivity analysis method (D.C.T.A.) of Claussen[15]. It may be used with

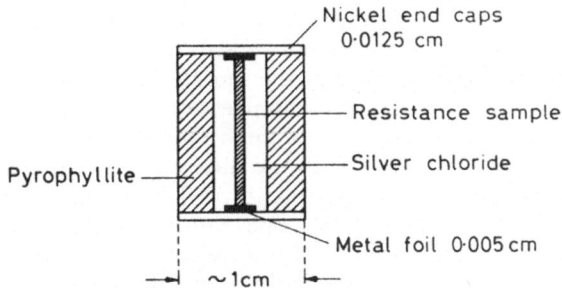

Figure 6.11a. Cell for resistance measurements at room temperature

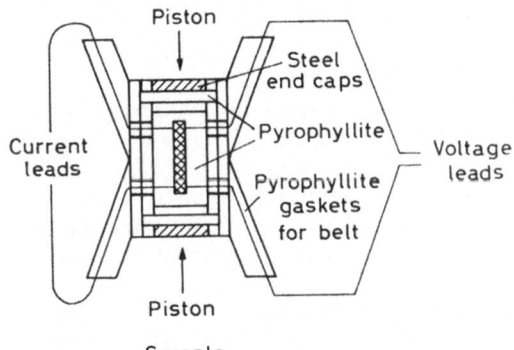

Figure 6.11b. High temperature resistance measurement cell (after Clougherty and Kaufman[14])

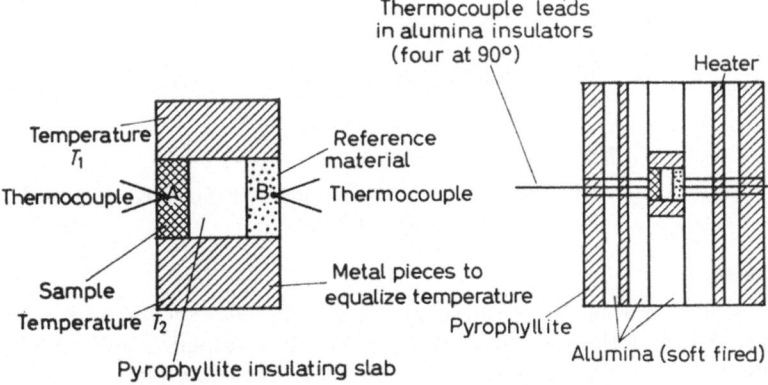

Figure 6.12. D.C.T.A. sample cell (after Claussen[15])

137

considerable advantage in the phase diagram analysis of, for example, iron alloys. With the arrangement shown, under normal conditions the temperature gradient across the sample and reference material is linear and the temperature difference between the points A and B is practically zero. If the sample undergoes a phase change there is a break in the linearity due to the different thermal conductivity of the two phases and this is accompanied by a finite temperature difference between A and B. The insulating pyrophyllite slab prevents this becoming smoothed out by contact with the reference material. An internal pressure calibration can be carried out if the reference material has a well known transition in the right region but sufficiently removed from any transitions in the sample. The reader is referred to the original papers for further details.

Differential Thermal Analysis

This technique has been used for phase-change determinations by Kennedy and his colleagues[16]. It is similar to the method of Claussen in that the thermocouple is placed in immediate contact with the sample and its reading compared with that of a similar one placed a small distance (few mms) away and not affected by any transitions (*Figure 6.13* shows part of the cell, the remainder follows that of *Figure 6.4*).

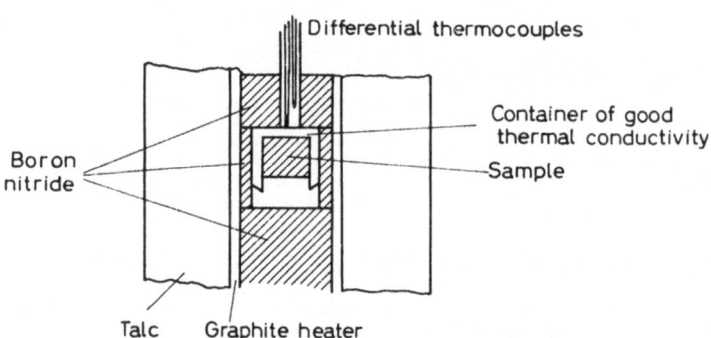

Figure 6.13. Differential thermal analysis sample cell (after Kennedy and Newton[16])

Synthesis

Cells suitable for synthesis are shown in *Figure 6.14a* and *b*. In the first case the sample itself is electrically conducting and may be heated by direct application of current. For example, the above G.E.C. patent (page 129) gives the following method for synthesizing

138

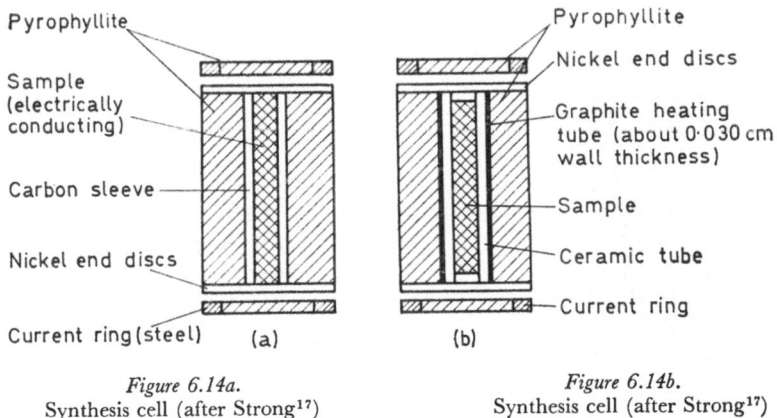

Figure 6.14a.
Synthesis cell (after Strong[17])

Figure 6.14b.
Synthesis cell (after Strong[17])

diamond. A cylindrical graphite tube is filled with a mixture of powdered graphite 15 parts, powdered iron 3 parts, manganese 1 part, and vanadium pentoxide 1 part, and is sealed with a graphite end plug. The application of 95 kb pressure and 1,700°C is then required to produce the graphite to diamond transition. In the second case (*Figure 6.14b*), samples are contained in copper tubes and placed inside ceramic liners and heated by the normal graphite tube. By this latter method it is possible in general to use liquids as the starting phase.

Magnetic Measurements

The cell assembly shown in *Figure 6.15* has been used in a piston and cylinder apparatus by Jayaraman[18] and others to determine Curie temperatures as functions of pressure. The magnetic sample is in the shape of a toroid and forms the core of a transformer. To take an example, a toroid of gadolinium 0·75 cm o.d. and 0·48 cm i.d. has two sets of 9 turns of 0·03 cm copper wire as primary and secondary coils. Currents of the order of 0·25 A at 1,000 c/s are passed through the primary and the sudden change in voltage in the secondary as the permeability of the core changes at the Curie point is noted. The ceramic tube has six bores to carry two thermocouple leads as well as the four transformer leads. Silver chloride and talc are used in the high-pressure cell and since usually Curie points are below room temperature, cooling fluids such as liquid nitrogen are passed through coils attached to the main piston and cylinder assembly.

139

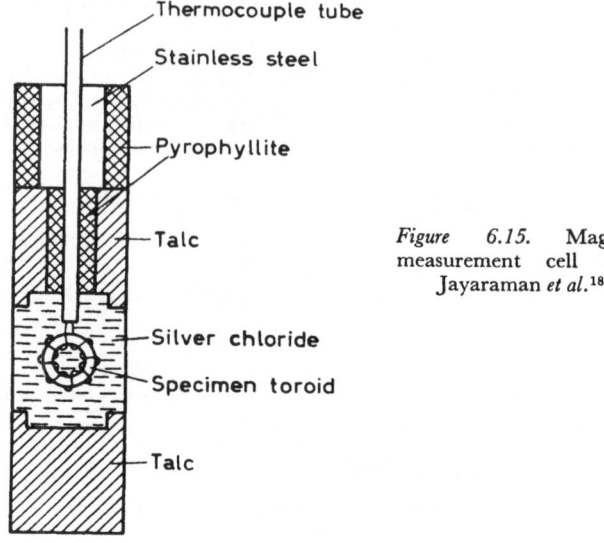

Figure 6.15. Magnetic measurement cell (after Jayaraman *et al.*[18])

Summary

The general behaviour[10,20] of the above cells is governed by many factors, the most important of which are temperature and pressure distribution throughout the volume. Several investigations[19,20] have been made by putting sensing elements in appropriate places but the results are usually only applicable in certain well defined cases.

REFERENCES

[1] Bridgman, P. W. *Proc. Roy. Soc.* 1950, **203A,** 1.
[2] Hall, H. T. *Rev. sci. Instrum.* 1958, **29,** 267.
[3] Christiansen, E. B., Kistler, S. S. and Gogarty, W. B. *Rev. sci. Instrum.* 1961, **32,** 775.
[4] Boyd, F. R. and England, J. L. *J. Geophys. Res.* 1960, **65,** 741.
[5] Boyd, F. R. *Modern Very High Pressure Techniques,* p. 151. Ed. R. H. Wentorf. Butterworths, London, 1962.
[6] Giardini, A. A., Tydings, J. E. and Levin, S. B. *Amer. Mineral.* 1960, **45,** 217.
[7] Bradley, R. S., Munro, D. C. and Whitfield, M. *J. sci. Instrum.* 1965, **42,** 714.
[8] Hall, H. T. *Rev. sci. Instrum.* 1959, **31,** 125.
[9] Bradbury, E. J. *High Pressure Measurement,* p. 186. Eds. A. A. Giardini and E. C. Lloyd. Butterworths, London, 1963.
[10] Bundy, F. P. *J. chem. Phys.* 1963, **38,** 631.
[11] Wilson, W. B. *Rev. sci. Instrum.* 1960, **31,** 331.

REFERENCES

[12] Daniels, W. B. and Jones, M. T. *Rev. sci. Instrum.* 1961, **32**, 885.
[13] Young, A. P., Robbins, P. B. and Schwartz, C. M. *High Pressure Measurement*, p. 262. Eds. A. A. Giardini and E. C. Lloyd. Butterworths, London, 1963.
[14] Clougherty, E. V. and Kaufman, I.. *High Pressure Measurement*, p. 152. Eds. A. A. Giardini and E. C. Lloyd. Butterworths, London, 1963.
[15] Claussen, W. F. *High Pressure Measurement*, p. 125. Eds. A. A. Giardini and E. C. Lloyd. Butterworths, London, 1963.
[16] Kennedy, G. C. and Newton, R. C. *Solids Under Pressure*, p. 163. Eds. Paul and Warschauer. McGraw-Hill, New York, 1963.
[17] Strong, H. M. *Modern Very High Pressure Techniques*, p. 93. Ed. R. H. Wentorf. Butterworths, London, 1962.
[18] Robinson, L. B., Milstein, F and Jayaraman, A. *Phys. Rev.* 1964, **134**, A187.
[19] Cohen, L. H., Klement, W. and Kennedy, G. C. *J. Phys. Chem Solids*, 1966, **27**, 179.
[20] La Mori, P. N. *High Pressure Measurement*, p. 321. Eds. A. A. Giardini and E. C. Lloyd. Butterworths, London, 1963.

7

MISCELLANEOUS METHODS

Many of the devices described in this chapter have common origins with those already discussed in Chapters 3 to 6 and only the factors influencing special applications will be given in the following sections.

HIGH-PRESSURE LOW-TEMPERATURE DEVICES

The application of high pressures to materials at temperatures lower than 100°K involves the development of several new techniques and materials. Some of the properties which have been studied under these extreme conditions include PVT measurements in solidified rare gases and alkali metals; superconductivity which arises from an electron lattice interaction of a particular type and is hence dependent on lattice spacing; and general physical properties of metals, for example bulk transport properties and Fermi-surface contours. The maximum pressures reached at liquid helium temperatures are an order of magnitude down on those obtainable at room temperature. Generally speaking it is possible to make accurate measurements to 27 kb under approximately hydrostatic conditions and to 45 kb with considerably less uniformity of pressure.

The requirements for generating high pressure in a low-temperature system are a suitable material to withstand the stresses involved, a pressure-transmitting medium which remains soft and preferably has good electrical insulating and chemical inertia properties, and a sealing mechanism. The most commonly used materials for pressure vessels are beryllium–copper alloys and stainless steels, which have the added advantages of being practically non-magnetic. The usual low carbon alloy steels like EN 26 which are ductile at normal temperatures become very susceptible to brittle fracture below certain transition temperatures. These vary with carbon and nickel content, heat treatment and grain structure but generally speaking the lower limit is about 150°K[1]. As a rule face-centred cubic structures tend to show no loss in ductility at low temperatures in contrast to the body-centred structures of which EN 25, 26 steels are examples. Austenitic stainless steels have face-centred cubic structures and have no ductile to brittle transition and can be used therefore at the lowest temperatures. One of the most commonly used types is EN 58 which is approximately 18 per cent chromium, 8 per cent nickel 0·08 per cent carbon (see Chapter 2 and Appendix A).

The use of beryllium–copper is discussed in the section on non-magnetic pressure vessels on page 154. Like austenitic stainless steel it does not lose its ductility at low temperatures and in fact becomes slightly stronger.

Tungsten carbide is comparatively brittle but has been employed successfully in piston and cylinder apparatus to 20 kb at liquid helium temperatures. Any piston or anvil or loading pad which is subject only to compression can be made from the usual materials.

In addition there is the range of maraging steels which retain strength and ductility at low temperatures but have not been used very extensively to date. For pressures of a few kilobars ceramics and single-crystal sapphires have been employed.

Although the cost of liquid helium is reasonably low at the present time it is advantageous to bear in mind that the smaller the scale of the high-pressure apparatus the less the cost of running an experiment since large quantities of liquid helium are usually involved. The consideration of size is also important with beryllium–copper vessels since smaller bulk pieces have less chance of containing blow holes of a size likely to cause rupture. Most low-temperature high-pressure apparatus does not involve support ring structures of the type described in the last chapter for these reasons.

The choice of a pressure medium for use at low temperatures is extremely limited. For example at 4·2°K only helium 3 and helium 4 are still fluid under one atmosphere pressure and these solidify at pressures of less than 200 b at this temperature. For most purposes, therefore, only solid transmitting media can be considered and the difficulty is to find one which provides the nearest approximation to hydrostatic conditions. The media used under normal temperature conditions such as silver chloride, pyrophyllite, boron nitride etc., are eliminated at low temperatures because of their extreme brittleness. (Bowen[2] has reported successful measurements on the effect of pressure on superconducting transitions using silver chloride but other authors have not found it so reliable and in all probability it is a question of the required degree of hydrostaticity.) Indium is fairly plastic at helium temperatures and is used widely but its electrical conducting properties are a disadvantage.

The weak binding forces in solid helium and solid hydrogen are an indication that these should be good hydrostatic pressure transmitting media. The difficulty in using them lies in the detrimental

143

effects they have on many materials normally used in the manufacture of pressure vessels. The embrittlement of steels by hydrogen is well known[3] and helium tends to go through 'holes' which normally do not exist for more conventional media. These difficulties can be overcome and pressure above 10 kb at temperatures down to 2°K have been generated using them.

There are two clearly defined types of high-pressure/low-temperature apparatus—those in which pressure is fixed at room temperature and a clamping device used to retain the pressure during subsequent cooling, and those in which the pressure is variable at low temperature by connection to a room temperature press. Both types make use of Bridgman anvils and piston and cylinder configurations for the high-pressure cells. Pressure calibrations and sealing methods will be considered individually below.

PRE-COMPRESSION DEVICES

The Ice Bomb Method of Lasarew and Kan[4]

One of the first and simplest forms of apparatus for generating pressure at low temperatures uses the phenomenon of the ice I–III, water triple point (Bridgman[5]). A beryllium–copper cell is filled with water, sealed off and cooled to 251°K (the triple point) (*Figure 7.1*). A pressure of about 2 kb is obtained and further

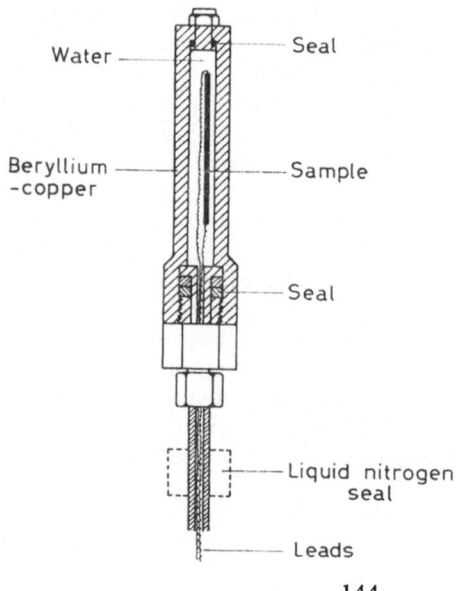

Water — Seal

Beryllium -copper — Sample

Seal

Liquid nitrogen seal

Leads

Figure 7.1. Ice bomb apparatus (after Lasarew and Kan[4])

cooling down to 4·2°K reduces this to about 1·7 kb due to differential thermal contractions. A more sophisticated version is shown in *Figure 7.2* which was first described by Brandt and Ginsburg[6]. In

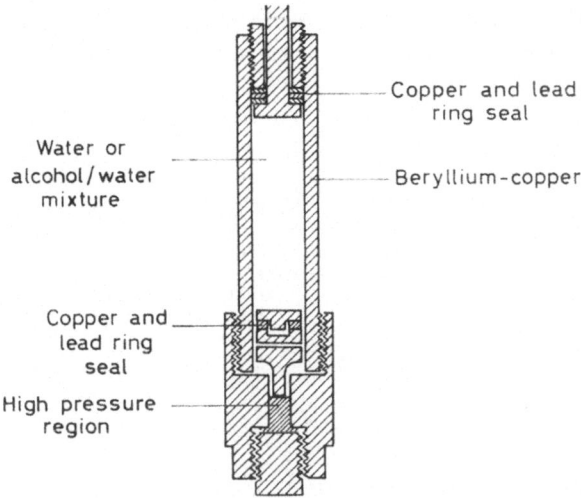

Figure 7.2. Ice bomb intensifier of Brandt and Ginsburg[6]
(not to scale)

this the ice I–III transition is used in the low-pressure end of an intensifier and with a multiplication factor of about 17, a theoretical pressure of 30 kb can be generated in the piston and cylinder high-pressure cell. Bridgman unsupported area seals are used with lead and copper sandwich packing. Although the pressure is fixed once the device has been loaded and cooled a variety of pressures between 0 and 30 kb can be obtained with alcohol and water mixtures in varying proportions. (The reader is referred to the original paper for a more detailed account of this.) Samples approximately 3 mm diameter and 3 mm long are used and are surrounded by a lubricant of thin cigarette paper impregnated with graphite. The bulk of the apparatus is constructed with hardened beryllium–copper (R.C. 40). The lower cylinder is prestressed to pressures above those used in the experiments. Pressure calibrations may be made using published data on the pressure dependence of the super-conducting transition in tin (see page 153) and the degree of hydro-staticity is inferred from the sharpness of this or other transitions (Brandt and Ginsburg estimate the maximum departure from hydrostaticity at 2–3 per cent).

145

Mechanical Locking Method

In the second precompression type the pressure is locked in at room temperature by a mechanical method and its retention down to helium temperatures relies on the similarity of thermal contraction of the material of the pressure device and the sample. The original version was built by Chester and Jones[7] and up to 40 kb is generated between Bridgman anvils at 4·2°K (*Figure 7.3*). The

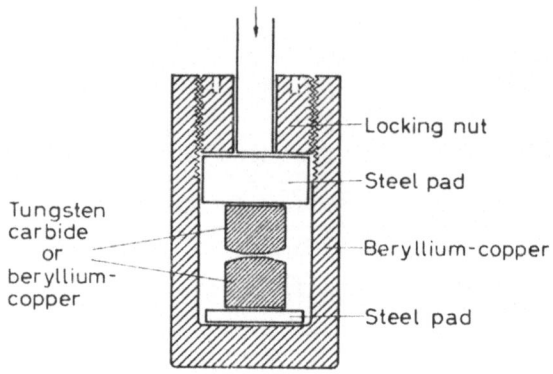

Figure 7.3. Low-temperature clamp (after Chester and Jones[7])

construction material is beryllium–copper although other materials can be used. The sample is compressed at room temperature between the anvils by means of a press. At the required load a locking nut is tightened onto the top face of the upper anvil as shown and the unit is then under a state of tension in, the outer walls and in compression through the Bridgman anvils. For the highest pressure tungsten carbide is used as the anvil material. The pressure is calculated from the force over area as indicated by a strain gauge and correcting for friction. The accuracy is between 5 and 10 per cent at 40 kb and the loss in pressure on cooling down to 4·2°K is less than 10 per cent. The main high-pressure requirement is that the tension in the outer cylinder should not exceed the yield stress for the material used, for example about 10 kb for beryllium–copper hardened to R.C. 40. Obviously the overall size should be kept to a minimum to avoid excessive use of liquid helium for cooling. This clamping device has been used with a tungsten carbide piston and cylinder and silver chloride pressure transmitting medium. The seals in this case were made with neoprene discs[2].

146

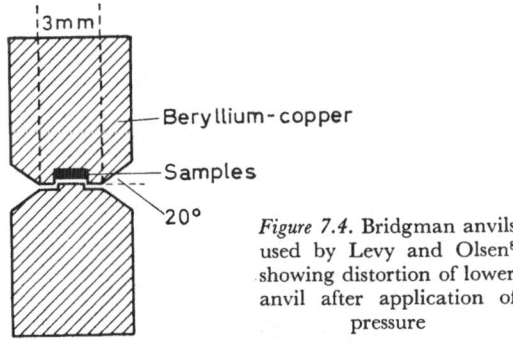

Figure 7.4. Bridgman anvils used by Levy and Olsen[8] showing distortion of lower anvil after application of pressure

A miniature version has been used down to liquid helium 3 temperatures by Levy and Olsen[8]. Due to the obvious smallness of a liquid helium 3 chamber the overall dimensions are reduced to about 1 by 2 cm and the sample is placed inside a cavity in one end of the upper Bridgman anvil together with a tin sample as a pressure calibrant (see page 153). The anvils are made from beryllium–copper (R.C. 40) and under load the bottom one distorts plastically to make a metal to metal seal to contain the sample and calibrant. The taper angle of the Bridgman anvils is between 10° and 20° with a flat area of 3 mm diameter. Pressures up to 21 kb may be generated and the width of the superconducting transition temperatures was found to be less than 0·005 in indium and less than 0·010°K in aluminium.

SOLID HELIUM PRESSURE DEVICE OF DUGDALE AND HULBERT[9]

In this method helium is used both as pressure transmitter and medium. It has been used mostly for PVT studies and electrical resistance measurements. The apparatus is illustrated in *Figure 7.5.* Helium gas at an appropriate pressure is solidified by cooling to a temperature which may be determined from the thermodynamic data given by Dugdale and Simon[10] (see *Table 7.1*). Subsequent

Table 7.1

Initial gas pressure	Solidification temperature, °K	Pressure at 2°K
3,000	28·3	2,300
2,500	25·2	1,900
2,000	21·9	1,500
1,500	18·2	1,100
1,000	14·1	730
500	9·1	350

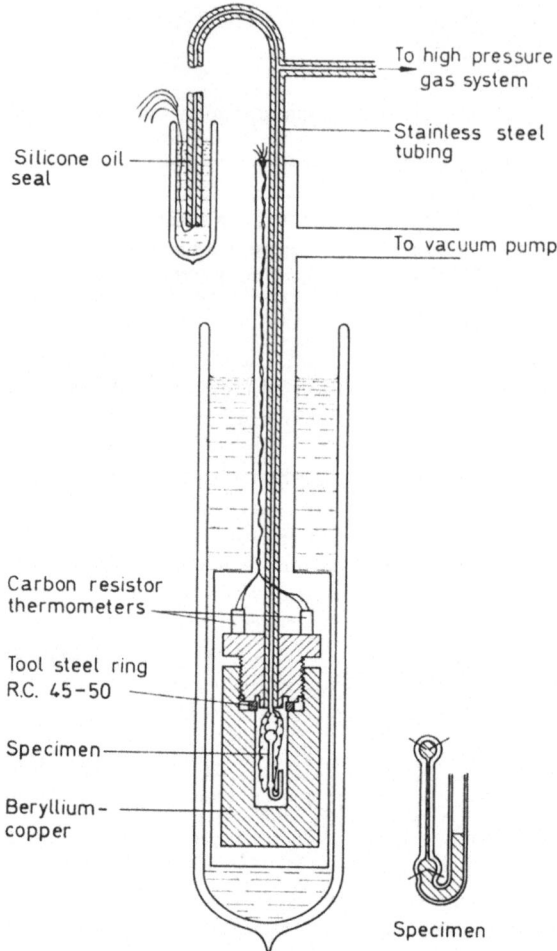

Figure 7.5. Solid helium high-pressure apparatus (after Dugdale and Hulbert[9])

cooling to 4·2°K after sealing the system off produces only a 10–20 per cent fall in pressure since the thermal expansion coefficient of solid helium falls rapidly with decreasing temperature. The exact pressure is determined from the equation of state (Dugdale and Simon[10]). In principle the operation is similar to that of the ice bomb but an important difference is that in order to change the pressure the helium has only to be heated to some tens of degrees

148

Kelvin compared with raising to room temperature in the former case. In the method of Dugdale and Hulbert only frozen-in pressures are used and the question of applying varying loads to solid helium (or solid hydrogen) will be considered in the next section.

The pressure vessel is constructed from beryllium–copper (R.C. 40) and is connected to a high-pressure input as shown. The seal between the cap and cylindrical container is made by tightening onto a hardened tool steel ring (R.C. 50) sufficiently to withstand up to 5 or 6 kb (see Chapter 3 for details of this type of seal). The high-pressure connection is through a stainless steel tube which is soft-soldered into the cap with an intermediate perforated copper disc. Electrical leads connected to a sample in the cylinder may be brought out through this tube and sealed in a frozen silicon oil bath as shown. The helium gas is pressurized using a conventional gas compressor and a Bourdon gauge serves as a monitor. For a precise knowledge of the pressure once the helium has solidified the temperature has to be kept steady to within a few milli-degrees and copper heating shields are used to protect the apparatus from room-temperature radiation. Normally an accuracy of 5 per cent can be obtained for the pressure. Dugdale and Hulbert soldered a copper braid half-way down the input high-pressure tubing to the inner can so that much of the heat influx is diverted from the high-pressure cell. The apparatus is cooled by introducing helium into the vacuum inter-space and the temperature measured by a carbon resistor or thermo-couple. From a safety point of view the sample volume is kept as low as possible in order to minimize the volume of compressed gas.

This method has been used to 8 kb by Shirber[11] for measurements on magneto-resistance in single crystal metals.

LOW-TEMPERATURE OPTICAL ABSORPTION CELL

The method of Dugdale and Hulbert has been employed in the construction of a high-pressure optical absorption cell for use at helium temperatures[12] (*Figure 7.6*). The cell has four windows and an input for high-pressure gas (only two windows are shown). The overall size is 5·5 cm diameter and 3·75 cm high which is in keeping with the aim of reducing the cost of cooling with liquid helium. The cell and window plugs are constructed from beryllium–copper alloy (R.C. 40) and the seals are Bridgman unsupported area types using bronze/indium/copper packing and beryllium–copper anti-extrusion rings. The apertures are sealed with single-crystal sapphires 6 mm diameter, 6 mm long in the Poulter arrangement (Chapter 3). [By replacing the latter with polished cylinders of

149

beryllium the apparatus can be used for x-ray diffraction experiments.] The window seals are held in place by a hardened steel support ring which is in turn held by an exit plug with a tapered aperture and may be removed in the usual manner. The pressure

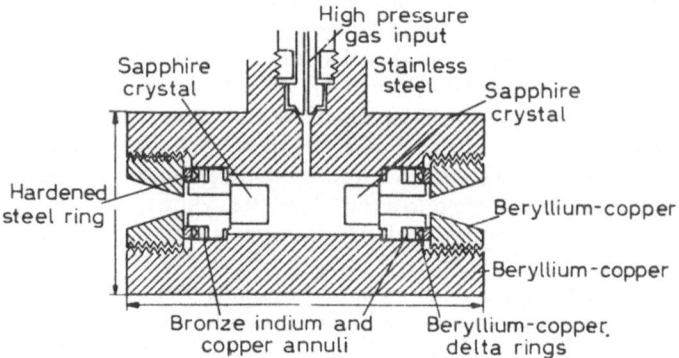

Figure 7.6. Low-temperature optical absorption cell (after Fitchen[12])

connection is through stainless steel tubing sealed to the cell by a standard fitting. Pressures up to 8 kb at low temperatures have been contained although there is some instability in the region 0–2 kb. Electrical leads are brought out using the method of Dugdale and Hulbert or through one or more of the window plugs. The cell is surrounded by a vacuum jacket to prevent any leaking gas boiling off helium. Cooling is through contact with a dewar containing helium and the whole unit sits in the tail of a cryostat approximately 9 cm o.d. The pressure is measured using the Dugdale and Simon equation of state data or by direct observation of an optical absorption band in an appropriate material. The minimum temperatures reached are slightly above those of liquid helium (i.e. 7 to 12°K) because of the radiation input through the optical windows in both the cell and the dewar.

PISTON DISPLACEMENT TYPES

Under this heading are included a number of devices which have been used extensively by Stewert[13] and Swenson[14] and others[15] for PVT and superconductivity experiments up to 27 kb under approximately hydrostatic conditions. The basic principle in this case is the external application of pressure to a cell at low temperature from an intensifier at room temperature via compression and tension cylinders or tie rods. A compromise is reached between the required strength of the device and the degree of heat influx which can be

150

tolerated. The basic form of the apparatus is illustrated in *Figure 7.7* where the compression generated in the inner cylinder wall by the intensifier is equally balanced by tension in a concentric outer cylinder. The high-pressure cell may be either piston and single-entry cylinder or a double-ended piston arrangement. The latter is probably the stronger because of weaknesses introduced in making blind hole cylinders but it doubles the sealing problem.

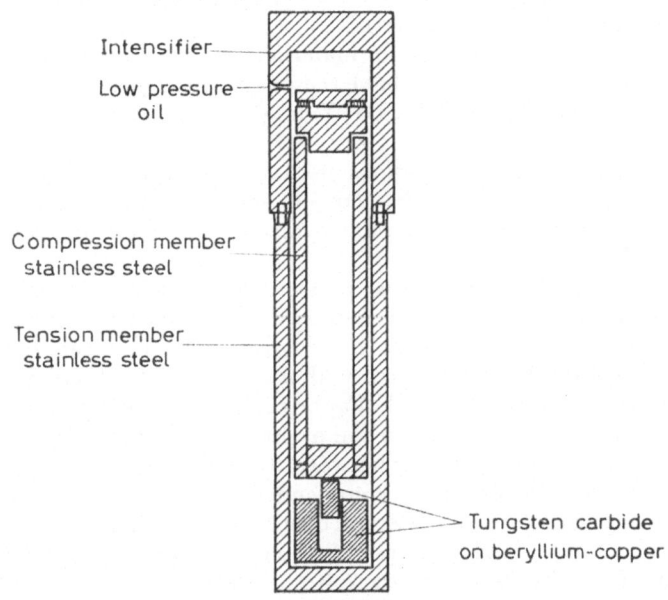

Intensifier

Low pressure oil

Compression member stainless steel

Tension member stainless steel

Tungsten carbide on beryllium-copper

Figure 7.7. Low-temperature piston displacement apparatus

The parameters of interest for a particular loading are:
(1) Compressive stress in the compression member.
(2) Tensile stress in the tension member.
(3) Buckling of the compression member.
(4) Heat influx through both members.
(5) Overall size and its relevance to liquid helium costs and size of the high-pressure chamber.

(1) and (2) are obvious, (3) arises because of the tendency of a comparatively long strut to buckle with a small lateral distortion. For a cylindrical shell the Euler criterion for buckling is given by

$$F = \frac{\pi^3}{4l^2} E \left(r_1{}^4 - r_2{}^4 \right)$$

151

where r_1 and r_2 are the radii, E is Young's modulus, l is the length, and F is the load. This condition should include also a factor to take into account the method of supporting the ends but here the most conservative value has been used[13]. It is seen from this formula that P is proportional to the inverse square of the length for given radii, which contrasts with the requirement of reducing heat influx by increasing the length. Therefore a compromise has to be reached together with the general considerations included in (5).

Stainless steels of the type EN 58 (18%Cr–8%Ni–0·08%C) are very suitable as construction materials for the compression and tension members since they have reasonably high yield strengths (in compression and tension) and Young's modulus and have low thermal conductivity.

The approximate dimensions used by Stewert are 5 cm o.d., 3·4 mm wall thickness and 3·75 cm o.d. and 3·4 mm wall respectively for the compression and tension cylinders and the total length is about 80 cm.

The hydraulic intensifier is pressurized from a standard hand pump and the load is conveniently measured with a manganin manometer. The pressure at the high-pressure end can only be calculated from force over area with a correction for friction unless an internal calibrant is used. Vaporization of helium due to heat influx from the upper part of the apparatus is reduced by a close-fitting dewar so that gas flow between the walls is turbulent (gaps of the order of 0·25 mm are recommended with the above sizes).

The high-pressure cell is constructed from hardened beryllium–copper, stainless steel or tungsten carbide for pressure in the 20 kb range. However, the first of these is not preferred for accurate PVT work because of its comparatively easy distortion.

The lower part of the apparatus used by Stewert[13] is shown in detail in *Figure 7.8*. A Bridgman mushroom plug seal is used with potassium metal packing which retains a reasonable degree of plasticity at low temperatures. In most cases either helium or hydrogen is used as the pressure-transmitting medium and is first condensed into the cell through a narrow side tube. Before compression the cylinder top is sealed by a thin copper sheet to contain the liquid. This is easily perforated as pressure is applied. The narrow side tube is sealed off in the early stages of the piston stroke and further compression up to 20 kb is then applied.

For experiments on samples embedded in the pressure medium electrical leads can be introduced through a bottom plug by means of the usual cone seals with pyrophyllite (Chapter 3) as the insulating material. These have been found to withstand up to 10 kb at 4·2°K[16].

A major problem with solid helium or hydrogen pressure media is the friction against the walls of the containing cylinder and the usual internal friction of a solid pressure medium. The former may be reduced by lining the walls with 0·125 mm thick indium foil and the latter can only be eliminated by careful experimentation.

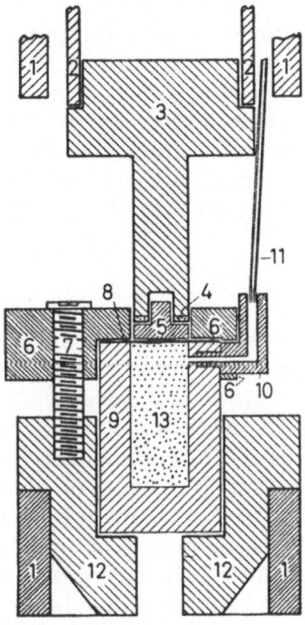

Figure 7.8. High-pressure piston and cylinder used by Stewart[18] in the press shown in *Figure 7.7*

1. Tension member
2. Compression member
3. Piston
4. Potassium, packing
5. 'Mushroom' plug
6, 7. Sample holder cap with screws for attachment
8. Copper sealing foil
9. Pressure cylinder (not to scale)
10, 11. Filling capillary and attachment to pressure cylinder
12. Sample holder support (attached to tension member)
13. Solidified gas sample

Pressure Calibration

The simplest method is to calculate the force over area with corrections for friction but this is not more than 5 to 10 per cent accurate.

The variation with pressure of superconducting transition temperatures and critical fields have been measured by many people for a large number of metals (see Bowen[2]). A manometer using the transition temperature for pure tin crystals may be used to give the pressure accurately to within a few per cent over the range 0–10 kb. The following equation is derived from the work of Jennings and Swenson[17] for tin;

$$T_c = 3 \cdot 732 - 4 \cdot 88 \times 10^{-5} P + 3 \cdot 8 \times 10^{-10} P^2 \quad (P \text{ in bars})$$

This account by no means exhausts the published work on low-temperature apparatus and the reader is referred to the review

153

articles by Brandt and Ginsburg[18] and by Levy and Olsen[19] for more information.

MAGNETIC RESONANCE EXPERIMENTS AT HIGH PRESSURE

Magnetic resonance experiments normally require the application of an external magnetic field to the sample under investigation, the exception to this being of course with ferro-, antiferro- and ferri-magnetic substances. The basic requirement therefore for a high-pressure container is a construction material which is either non-magnetic or only weakly magnetic. In addition the apparatus has to be on a small scale so that it can be placed between the poles of the normal type of resonance magnet.

Non-magnetic Pressure Vessels

The construction and use of these pressure vessels in the hydro-static pressure range has been reviewed by Paul *et al.*[20] and only the basic facts will be given here. The two most suitable materials are beryllium–copper and austenitic stainless steel. Cylinders of these may withstand up to 20 kb after initial cold working. Beryllium–copper alloy has usually the composition by weight of about 2 per cent beryllium, 0·26 per cent cobalt and the remainder copper and is obtainable from manufacturers in a fairly soft condition after solution annealing at 780–800°C. It is hardened by the process of precipitation by heating to around 350°C for up to two hours and quenching. (These figures are only a general guide and the instructions given by the manufacturers should be followed.) 10 kb proof stress at 0·1 per cent offset can be obtained (Chapter 2).

Cylinders of beryllium–copper can be cold worked by applying internal pressures greater than 10 kb with a soft metal such as indium as a medium. It is important when machining the hardened metal that the temperatures should not approach the precipitation temperature.

Austenitic stainless steels (EN 58 E) are used as alternatives to beryllium–copper and may be cold worked to give strength up to 14 kb. Although both materials are used to 20 kb and beyond, above 14 kb there is considerable plastic flow which increases the sealing problems.

To summarize, stainless steel is used at R.C. 32 and is easily machined in this state, in addition it has low heat conductivity and may be expected to be free from flaws in the raw state. Beryllium—

copper is hardened to R.C. 40 by a more easily carried out process than in the case of stainless steel but may well contain initial flaws especially in large pieces which must be detected by ultrasonic methods before using in high-pressure apparatus.

NUCLEAR MAGNETIC RESONANCE

A high-pressure cell suitable for studying nuclear magnetic resonance up to 10 kb is shown in *Figure 7.9* and was first constructed by Paul *et al.*[20]. The high-pressure aspects are conventional with respect to the plugs and the electrode seals (Chapter 3).

The Drickamer supported anvil apparatus may be modified to make n.m.r. studies to 80 kb[21]. This is based on the tapered anvil with support which was described in Chapter 4. The supporting material is boron nitride in this case since pyrophyllite contains protons and much of n.m.r. work involves proton resonance. The apparatus is shown in *Figure 7.10* and is constructed mainly of hardened beryllium–copper. The usual intensifier is used and the bottom end is designed to go between the pole pieces of a magnet. The high-pressure anvils are made of chrome oxide which has the advantage of being both an insulator and non-magnetic thereby reducing the radio-frequency energy losses. The flats are 6 mm diameter with a 45° taper. The size of the resonance coil depends on the inductance required and therefore on the sample within it. For proton work fillers such as epoxy resins cannot be used and hence the copper turns are made of the thickest gauge which can fit into the available space. One lead of the coil goes to ground while the other passes through a 0·40 mm hole and is insulated with teflon. The sample cavity is about 0·2 mm in height and the overall width at the bottom end is about 6·5 cm. The usual techniques with weak signals are used in the n.m.r. spectrometer and it is possible to make spin echo measurements.

Nuclear Magnetic Resonance in Ferro-magnetic Materials

Resonance in ferromagnetic materials generally does not require the application of external fields. The type of high-pressure equipment which has been described in previous chapters may be used. For experiments on iron-57, Litster and Benedek[22] have adopted the belt apparatus in the form described in Chapter 6. The cell assembly in this case is shown in *Figure 7.11*. The pressure medium is silver chloride containing powdered iron in which is embedded a five-turn coil of copper wire. A mixture of iron and silver chloride in the

155

ratio 1 to 14 by weight is compacted with the coil to 7 kb in a die and then machined to a cylindrical form 1·27 cm long and 0·6 cm diameter.

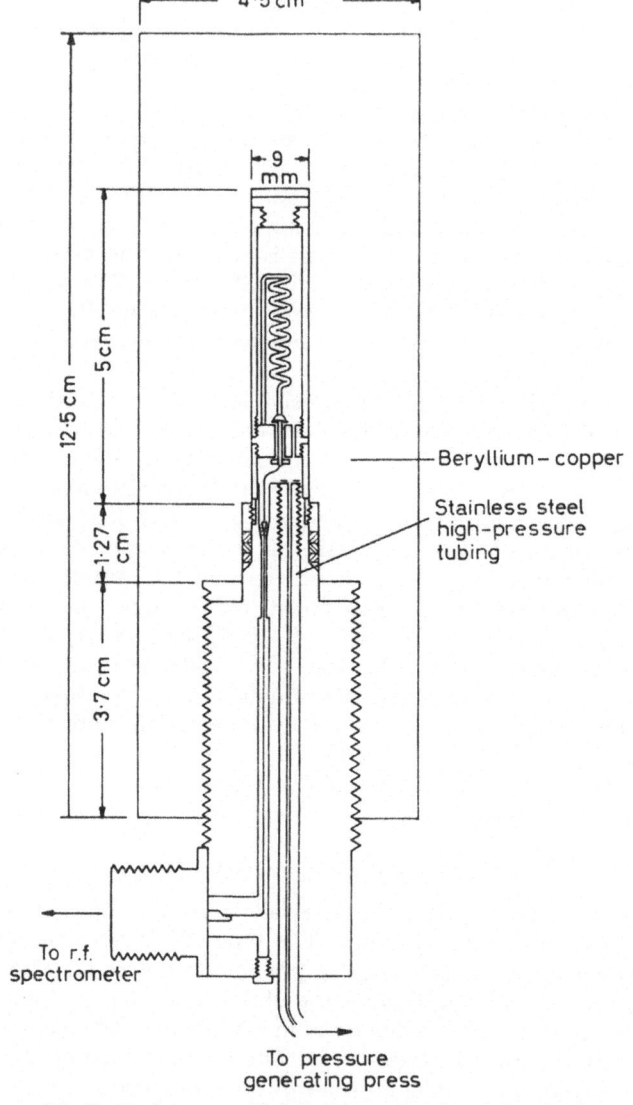

Figure 7.9. Radio-frequency high-pressure assembly (after Paul *et al.*[20])

156

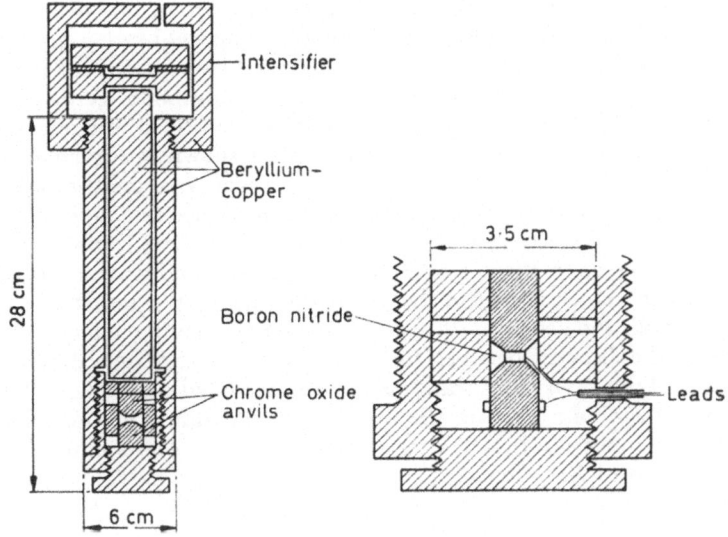

Figure 7.10. High pressure n.m.r. cell (after Cleron *et al.*[21])

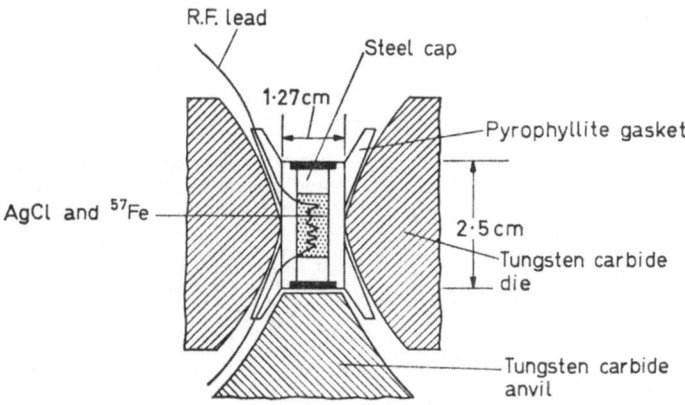

Figure 7.11. High pressure. nm.r. cell for belt apparatus (after Litster and Benedek[22])

ELECTRON SPIN RESONANCE

As far as high-pressure experiments are concerned the main difference between e.s.r. and n.m.r. measurements is in the coupling in of the electromagnetic energy. In the case of e.s.r. the normal

157

conductors cannot be used and a waveguide system has to be incorporated into the high-pressure device. This requires a correct impedance match and the usual conical and step entry into high-pressure cylinders creates reflection and multimode generation problems. The three types which have been developed to overcome these difficulties in hydrostatic apparatus are shown in *Figures 7.12, 7.13* and *7.14*. In the first[23] the microwave energy enters through a 50-ohm connector followed by a 50-ohm impedance line tapering to 0·25 cm i.d. which then widens to 0·6 cm with a beryllium–copper cone seating on a mica washer as the centre conductor. A brass microwave cavity is press-fitted onto the end of the insert and is coaxial with the centre stud. Correct matching may be achieved by altering the distance between this stub and the cone. The remaining volume of the cavity and transmission line are filled with teflon. The loaded Q is about 500 and the upper frequency limit 10 Gc/s. The pressure connection is of the usual type using petroleum ether as a medium up to 10 or 15 kb.

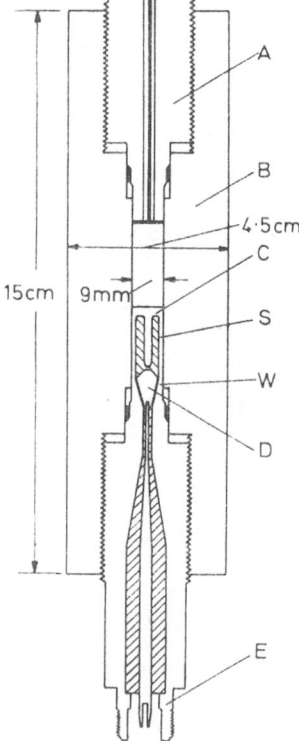

Figure 7.12. High-pressure bomb and microwave transmission plug (after Walsh and Bloembergen[23])

A. Pressure feed plug
B. Be–Cu bomb
C. Microwave cavity
D. Sealing cone
E. Type N connector
S. Sample position
W. Synthetic mica washer

A disadvantage of this microwave connection is that under load the compressible seal using the beryllium–copper cone changes dimension under high pressure and causes difficulties with mismatching. *Figure 7.13a* and *b* illustrates an alternative type in which single-crystal sapphires are used both as dielectric filler for the cylindrical input waveguide and as sealing cones for the high-pressure end. The two crystals are cut so that they abut firmly to eliminate any air gap which would introduce a high impedance. The larger insert is wedge-shaped so that there is a progressive change in impedence along the waveguide. The included cone angle of the smaller crystal is 16° following the usual design and it is found that 10 kb pressure may be contained. The cavity is mounted as shown.

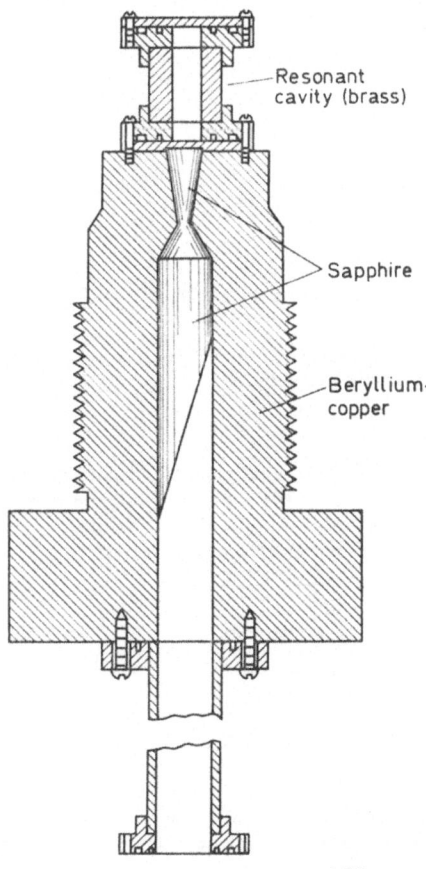

Resonant
cavity (brass)

Sapphire

Beryllium–
copper

Figure 7.13a. High-pressure plug showing two opposing tapered sections of circular guide drilled along its axis. The resonant cavity is mounted by screws on top of the plug

159

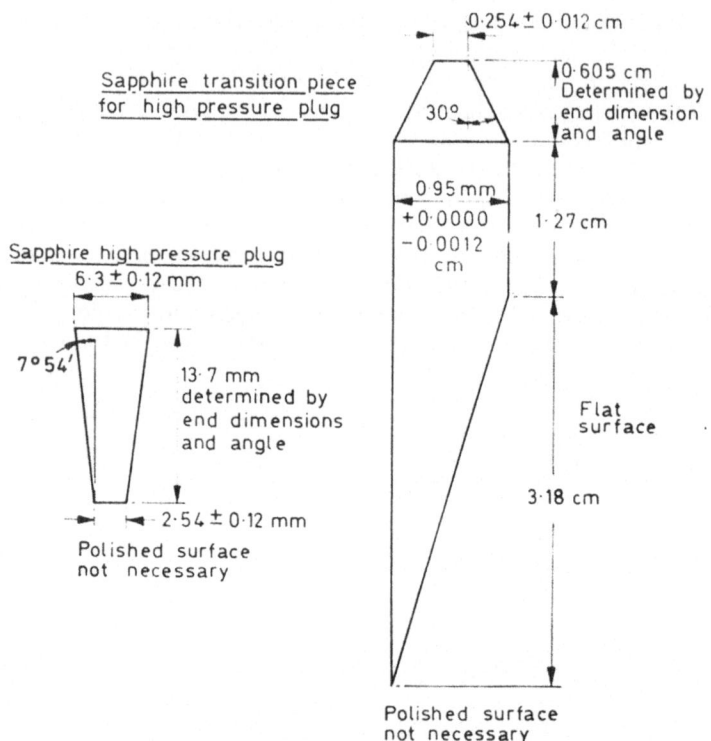

Figure 7.13b. Dimensions of the sapphire inserts used in the microwave transition (after Lawson and Smith[24])

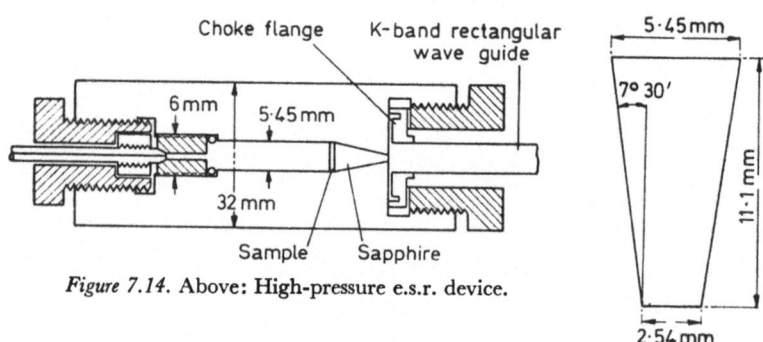

Figure 7.14. Above: High-pressure e.s.r. device.

Right: Dimensions of the sapphire cavity. The length and taper determine the resonant frequency (after Goodrich *et al.*[25])

160

In both the above devices there is the disadvantage that under pressure both the cavity and pressure medium change their dielectric properties. In addition, the coaxial method is not suitable for the highest frequencies. *Figure 7.14* illustrates an apparatus in which a sapphire crystal cone is shaped appropriately so that it forms the resonant cavity with the sample placed immediately at the high-pressure end. The sapphire has to be formed accurately and perfectly in order to get correct matching. The size shown in *Figure 7.14* is appropriate for 22·5 Gc/s. The Q value depends on the degree of parallelism and is of the order of 500. It is found that the resonant frequency changes by less than 2 Mc/kb.

Although sapphire inserts have been used successfully by the above authors, in other laboratories they have broken, particularly on unloading, at quite low pressures (a few kb).

For much higher pressures a Bridgman anvil apparatus has been used by Gardner et al.[26]. The lower and upper anvils are made from cold and hot pressed alumina respectively *(Figure 7.15)*. The

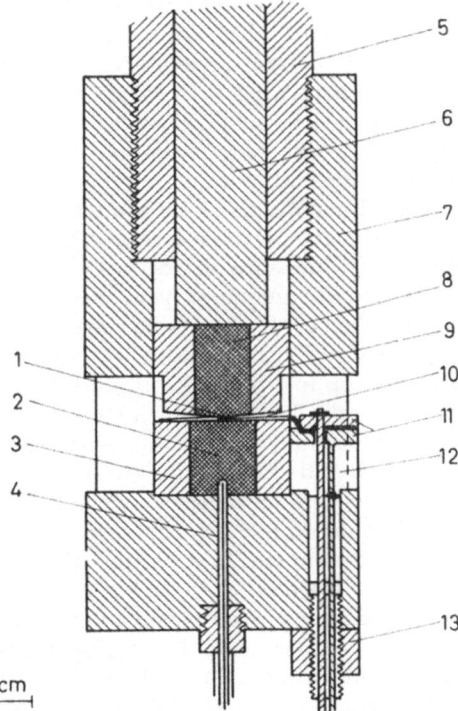

Figure 7.15. High pressure device for e.s.r. measurements to 60 kb (after Gardner et al.[26]). Schematic diagram of high-pressure head.

1. Sample cell
2. Combined pressure anvil and microwave cavity
3. Beryllium–copper binding ring for pressure anvil
4. Coaxial line for coupling microwaves
5. Beryllium–copper sleeve for coupling to hydraulic press
6. Beryllium–copper pressure ram
7. Beryllium–copper pressure head
8. Bridgman type tapered pressure anvil
9. Beryllium–copper binding ring for pressure anvil
10. Silver modulation strip
11. Modulation clamp
12. Plastic insulating support for modulation clamp
13. Locking nut for modulation clamp

1 cm

161

upper anvil is shaped with flats and the lower one is coated with silver and acts as the cavity. The alumina is reinforced with beryllium–copper rings with 0·05 mm interference on a 3/4° taper. The upper anvil is connected to conventional intensifier and pressures up to 60 kb at temperatures down to 100°K can be generated between the anvils. Cold pressed alumina is weaker but purer and therefore less lossy than the hot pressed variety. The microwave power is coupled in by a straight probe placed midway between the axis and the wall of the lower cavity. Correct matching is achieved by varying the distance the probe enters the anvil.

The reader is referred to the original papers for more details on the magnetic resonance measurements.

NEUTRON DIFFRACTION AT HIGH PRESSURES

The requirements of a neutron diffraction experiment at high pressure can be summarized under the following headings.

(a) A material strong enough to hold high pressures yet transparent to a neutron beam about 1·2 Å wavelength.

(b) A pressure medium which is hydrostatic and again transparent.

(c) A sample volume of the order of a few cm³.

(a) and (b) are obvious, (c) arises because of the comparative weakness compared with x-rays of present-day sources of thermal neutrons. Unless single crystals are used a large signal-to-noise ratio in a diffraction pattern can be obtained only using a beam of the order of several millimetres diameter, which in turn requires a similar size polycrystalline sample. Large in and out ports for the neutron beam are therefore necessary which means that the most useful high-pressure device is the piston and cylinder which has the added advantage of being axially symmetrical.

The choice of a container material is narrowed down by the large absorption of neutrons by most high-strength materials. For pressures up to 1 or 2 kb it is possible to use thin-walled steel vessels but up to the 10 kb region the choice is reduced to either synthetic sapphire crystals or a number of alloys of aluminium, titanium, zirconium and zinc. At the National Physical Laboratory a number of single-crystal sapphire cylinders 2·5 cm long and 2·5 cm diameter with k ratios up to 5 have consistently failed at less than one quarter of the ultimate tensile stress and although this may be due to some extent to incorrect relative orientations of the c axis and the cylinder bore it has not proved possible to use them above 1 or 2 kb. Single-crystal containers have the considerable advantage of a background diffraction pattern with a few large narrow peaks and relatively large sections of the 2θ angle with very low background and hence admirably suitable for showing up weak diffraction peaks from magnetic samples.

Smith et al.[27] have used cylinders of an alloy of aluminium with 7 per cent zinc with a nominal tensile strength of 5 kb. Considerable plastic deformation was tolerated in using them to 10 kb. The background diffraction of this material is free from large irregularities up to a 2θ value of 27°. Pistons of a much stronger material such as K9 tool steel are used and a suitable pressure medium is carbon disulphide which is relatively transparent to low-energy neutrons, containing no hydrogen as do most oils and having a solidification pressure of 12 kb at room temperature. The piston and cylinder are fitted into an intensifier arrangement as shown in *Figure 7.16* and

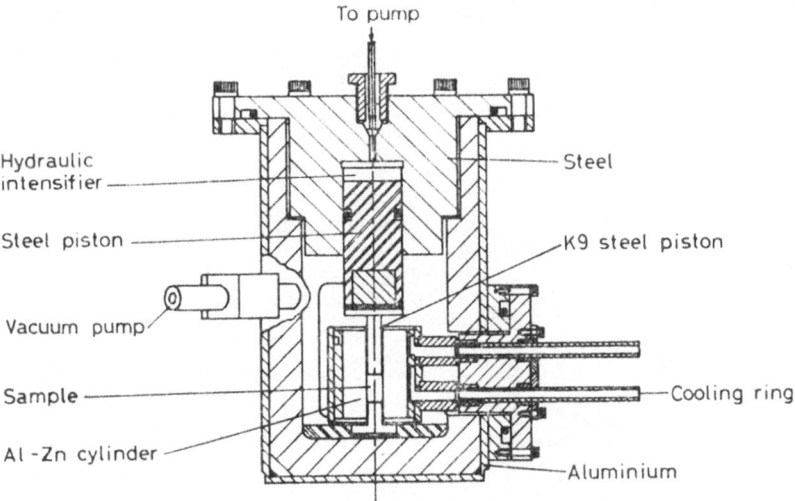

Figure 7.16. High-pressure neutron diffraction apparatus (after Smith et al.[27])

the whole unit is in an aluminium shell and can be heated to 250°C or cooled to −180°C. It is found that if liquid carbon disulphide is cooled under pressure through the solidification transition reasonably hydrostatic conditions are maintained. The heating and cooling jackets have windows in appropriate places to transmit the neutron beam and the unit is evacuated to 10^{-3} Torr. Pressure measurement is carried out using either manganin resistance gauges or by recording the oil pressure in the intensifier with a Bourdon gauge and making friction corrections by cycling the pressure. It is a feature of both neutron and x-ray diffraction experiments in crystals that relative volume changes are more relevant than a knowledge of the actual pressure. This is of course easily determined for any temperature and pressure from the position of the diffraction peaks.

The above device has been used with a neutron diffraction spectrometer at A.E.R.E., Harwell, to study the helicoidal antiferromagnetic structure of Au_2Mu[27].

Figure 7.17 illustrates two piston and cylinder devices which have

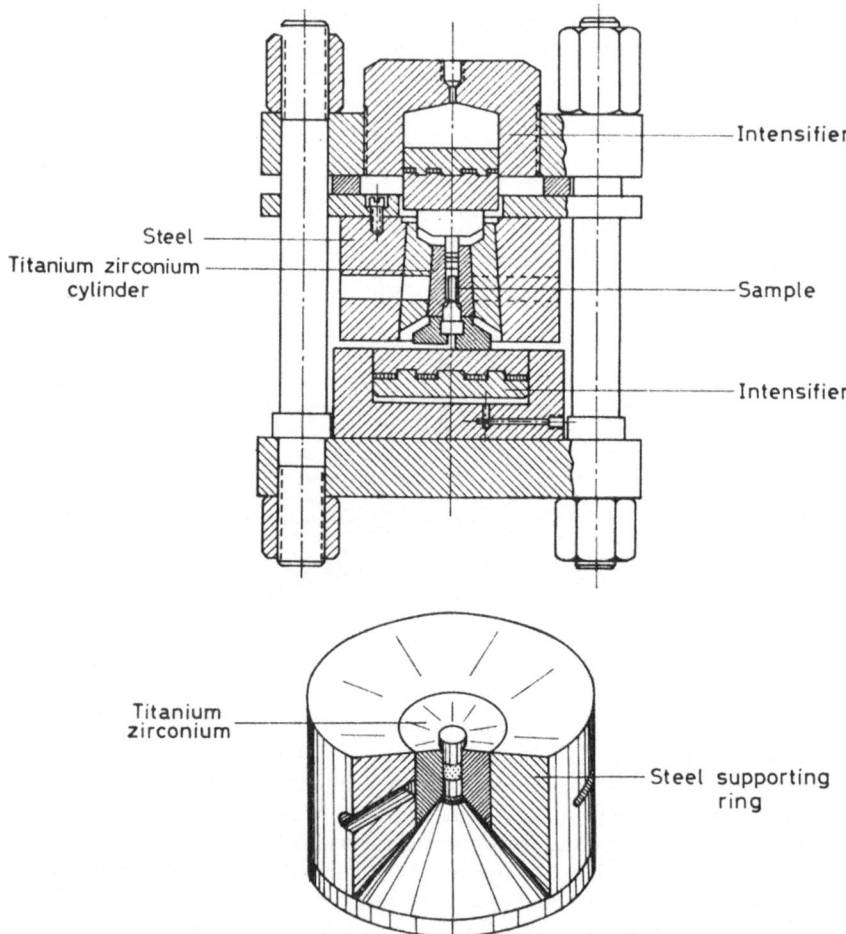

Figure 7.17. High-pressure neutron diffraction device (after Litvin and Ponyatovskii[28])

been used by Litvin[28] for neutron diffraction experiments. The inner cylinders are 50/50 titanium zirconium alloy which is supported by steel cylinders in which slits are cut to allow the transmission of the neutron beam. In the first arrangement which is used for single crystal studies the Bridgman method of forcing the inner cylinder into the outer one under successive loads is used (see page 49). The included angle of the cone shaped surfaces is 6°. Mushroom plug seals of rubber and copper are used and hydrostatic pressures up to 7 kb may be generated. In the second type which is used for polycrystalline studies the internal bore is some 30 per cent larger and the sample is compressed directly with copper or teflon liners 0·1 mm thick. With wall thicknesses up to 2·5 cm and shrunk-on steel supporting cylinders pressures up to 28 kb have been generated.

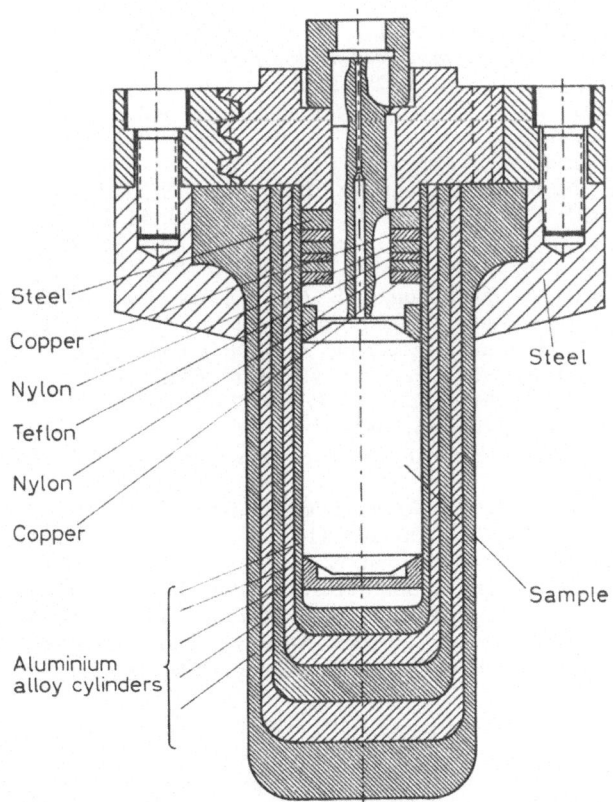

Steel

Copper

Nylon

Teflon

Nylon

Copper

Steel

Aluminium alloy cylinders

Sample

Figure 7.18. High-pressure neutron diffraction cell (after Lechner[29])

165

For inelastic neutron-scattering experiments at comparatively low pressures (less than 3 kb) Lechner[29] has used a compound cylinder construction with a total internal to external diameter of approximately two (*Figure 7.18*). The comparatively thin walls are more important for inelastic than elastic scattering experiments because of keeping background diffraction to a minimum. The material employed is an aluminium alloy (6·5 per cent zinc, 3 per cent magnesium and 2 per cent copper) and the five cylinders are heat-shrunk together with interference varying from 0·35 mm at the inner contact surface to 0·100 mm at the outermost one. The radii are in a geometrical series as is normally the rule with compound cylinder constructions.

REFERENCES

[1] Parker, C. M. and Sullivan, J. W. W. *Ind. Eng. Chem.* 1963, **55** (5), 18.
[2] Bowen, D. H. *Proceedings of the 5th International Conference on Low Temperature Physics and Chemistry*, p. 337. Madison, Wisconsin, 1958.
[3] Bridgman, P. W. *The Physics of High Pressure*. Bell, London, 1952.
[4] Lasarew, B. G. and Kan, L. S. *J. Phys. USSR*, 1944, **8**, 361.
[5] Bridgman, P. W. *J. chem. Phys.* 1937, **5**, 964.
[6] Brandt, N. B. and Ginsburg, N. I. *Sov. Phys. J.E.T.P.* 1961, **12**, 1082.
[7] Chester, P. F. and Jones, G. O. *Phil. Mag.* 1953, **44**, 1281.
[8] Levy, M. and Olsen, S. L. *Rev. sci. Instrum.* 1965, **36**, 233.
[9] Dugdale, J. S. and Hulbert, J. A. *Canad. J. Phys.* 1957, **35**, 720.
[10] Dugdale, J. S. and Simon, J. E. *Proc. Roy. Soc.* 1953, **A218**, 291.
[11] Shirber, J. E. *Physics of Solids at High Pressure*, p. 46. Eds. E. T. Tomizuka and K. M. Emrick. Academic Press, New York, 1965.
[12] Fitchen, D. B. *Rev. sci. Instrum.* 1963, **34** (6), 673.
[13] Stewert, J. W. *Modern Very High Pressure Techniques*. Ed. R. H. Wentorf. Butterworths, London, 1962.
[14] Swenson, C. A. *Progress in Very High Pressure Research*. Wiley, New York, 1961.
[15] Stajdohar, R. E. and Towle, L. C. *A.S.M.E. 64-WA/Pt-9*, 1964.
[16] Hatton, J. *Phys. Rev.* 1956, **103**, 1167.
[17] Jennings, L. D. and Swenson, C. A. *Phys. Rev.* 1958, **112**, 31.
[18] Brandt, N. B. and Ginsburg, N. I. *Sov. Phys. Uspekhi*, 1965, **8**, 202.
[19] Levy, M. and Olsen, J. L. *Physics of High Pressure and the Condensed Phase*, p. 525. Ed. A. van Itterbeek. North Holland. Amsterdam, 1965.
[20] Paul, W., Benedek, G. B. and Warschaur, D. M. *Rev. sci. Instrum.* 1959, **30**, (10), 874.
[21] Cleron, V., Coston, C. J., and Drickamer, H. G. *Rev. sci. Instrum* .1966, **37** (1), 68.
[22] Litster, J. D. and Benedek, G. B. *J. appl. Phys.* 1963, **34** (3), 688.
[23] Walsh, W. M. jr. and Bloembergen, N. *Phys. Rev.* 1957, **107**, 904.
[24] Lawson, A. W. and Smith, G. E. *Rev. sci. Instrum.* 1959, **30** (11), 904.

REFERENCES

[25] Goodrich, R. G., Everett, G. E. and Lawson, A. W. *Rev. sci. Instrum.* 1964, **35** (11), 1596.
[26] Gardner, J. H., Hill, M. W., Johansen, C., Larsen, D., Murri, W. and Nelson, M. *Rev. sci. Instrum.* 1963, **34** (a), 1043.
[27] Smith, F. A., Bradley, C. C. and Bacon, G. E. *J. Phys. Chem. Solids*, 1966, **27,** 925.
[28] Litvin, D. F. and Ponyatovskii, E. G. *Sov. Phys. Crystallogr.* 1966, **11** (2), 322.
[29] Lechner, R. *Rev. sci. Instrum.* 1966, **37** (11), 1534.

APPENDIX A

MATERIALS COMMONLY USED IN HIGH-PRESSURE APPARATUS

Alloy Steels (nickel chromium molybdenum)

Type	Principal alloying elements %						U.T.S.*
	C	Si	Mn	Ni	Cr	Mo	
EN 24	0·35/0·45	0·1/0·35	0·45/0·7	1·3/1·8	0·9/1·4	0·2/0·35	~ 10·5 kb
EN 25	0·27/0·35	0·1/0·35	0·5/0·7	2·3/2·8	0·5/0·8	0·4/0·7	~ 10·5
EN 26	0·36/0·44	0·1/0·35	0·4/0·7	2·3/2·8	0·5/0·8	0·4/0·7	~ 12
EN 30B	0·26/0·34	0·1/0·35	0·4/0·6	3·9/4·3	1·1/1·4	0·2/0·4	~ 15

Young's modulus ~ 1,950 kb

Linear thermal expansion coefficient 12–13 $\times$ 10^{-6}/°C.

* After following manufacturer's recommended heat treatment.

Tool Steels (Edgar Allen range)

Type	Principal alloying elements %									U.T.S.*
	C	Si	Mn	Cr	Mo	Co	V	W	Ti	
K 9	1·0		0·85	0·75				0·4		~ 23 kb
Double Six	1·9			12·5			0·5			~ 32 kb
Stag Major	0·8	0·3	0·7	5·0	0·5	11·5	1·5	21·0		~ 23 kb

Young's modulus ~ 2,100 kb (K 9 and Double Six)
~ 2,500 kb (Stag Major)

Linear thermal expansion coefficient 10^{-5}/°C.

* After following manufacturer's recommended heat treatment.

Maraging Steel. Rex 684 (Firth Brown, Ltd.)

Type	Principal alloying elements %							U.T.S.
Rex 648	C	Si	Mn	Ni	Mo	Co	Ti	
	0·17	0·1	0·03	18·1	4·8	7·5	0·6	~ 18 kb

Stainless Steel

Type	Principal alloying elements %					U.T.S.
	C	Si	Mn	Ni	Cr	
EN 58 E	0·08	0·2	2·0	8·0/ 11·0	17·5/ 20·0	Depends on cold working. 10 kb

Young's Modulus ~ 1,800 kb

Linear thermal expansion coefficient ~ $17 \times 10^{-6}/°C$.

Beryllium–copper

Cu-Be 250. 2 % Be, 0·26 % Co, remainder copper. (Telcon Ltd.)

U.T.S. ~ 10·5 kb (heat treated after hard rolled)

Young's modulus ~ 1,300 kb

Linear thermal expansion coefficient.

Tungsten Carbide

Composition % Cobalt binder	U.T.S. in tension	U.T.S. in compression	Young's modulus	Expansion coefficient
3	~ 22 kb	~ 54 kb	~ 7,700 kb	$2·8 \times 10^{-6}/°C$
6	~ 15 kb	~ 40 kb	~ 7,000 kb	$2·8 \times 10^{-6}/°C$

Non-metals

Hot-pressed alumina

U.T.S. in compression	$\sim$	32 kb
U.T.S. in tension	$\sim$	7 kb
Young's modulus	$\sim$	4,500 kb

Single-crystal sapphire

U.T.S. in compression	$\sim$	22 kb
U.T.S. in tension	$\sim$	7–4 kb
Young's modulus	$\sim$	4,000 kb

Boron carbide

U.T.S. in compression	$\sim$	30 kb
U.T.S. in tension	$\sim$	3·5 kb
Young's modulus	$\sim$	3,000 kb

APPENDIX B

SUPPLIERS OF MATERIALS AND EQUIPMENT IN THE UNITED KINGDOM

The following list is by no means a comprehensive summary but is included as a guide.

Alloy Steels
English Steel Corporation Ltd., River Don Works, Sheffield.
Hadfields Ltd., East Hecla Works, Sheffield.

Tool Steels
Edgar Allen Ltd., Sheffield, 9.
Jessops, Ltd., Sheffield, 3.
Richard Carr Ltd., Wadsley Bridge, Sheffield, 6.

Stainless Steel
Firth Brown, Ltd., Sheffield, 4.

Tungsten Carbide
Murex, Ltd., Rainham, Essex.
Metropolitan Vickers, Ltd., Manchester.

Beryllium–Copper
Telcon Metals, Ltd., Crawley, Sussex.
Beryllium and Copper Alloys, Ltd., Victoria St., London, S.W.1.

Sapphire (single crystals)
Salford Instrument Co., Heywood, Lancs.
Agate Products, Ltd., Merton Park, London, S.W.20.

Hot Pressure Alumina
Carborundum Co., Ltd., Trafford Park, Manchester.

Diamonds
For windows. Van Moppes, Ltd., Basingstoke.
Triefus Industries, Manor Royal, Crawley.

Anvils.   A. Shore & Sons, Waterloo Works, Waterloo Road, London, N.W.2.
Star Industrial Tools, Westfield Road, Birmingham, 14.

Pyrophyllite
  Wm. Sugg & Co., Ltd., Ranelagh Works, Chapter St., London, S.W.1.
  Schaefer Diectrics, Ltd., 30, Thompsons Land, Cambridge.

Talc
  Wm. Sugg & Co., Ltd., Ranelagh Works, Chapter St., London, S.W.1.

Small Hand-operated Oil Pumps ($\sim$ 1 kb)
  Tangye, Ltd., 12, Waterloo St., Glasgow, C.2.
  Blackhawk Applied Power Industries, Inc., (U.K.), 717, Tudor Estate, Abbey Rd., London, N.W.10.

Large-Scale Presses
  Avery, Ltd., Birmingham.
  Farnell, Ltd., North Mimms, Hatfield, Herts.
  Amsler, T. C. Howden, Ltd., Leamington Spa (U.K. Agents).
  Fielding & Platt, Ltd. Gloucester.

Small Hand-operated Press (*30 Tns*)
  Research & Industrial Instrument Co., London.

Piping, Valves etc. for Hydraulic Pressure Generation
  Tangye, Ltd., 12 Waterloo Road, Glasgow, C.2.
  Blackhawk. Applied Power Industries, Inc., (U.K.), 717 Tudor Estate, Abbey Road, London, N.W. 10.
  Pressure Products, Inc., Ltd., Shaw Lane, Glossop, Derbyshire.

The author is indebted to Mr. N. B. Owen of the National Physical Laboratory for the information in Appendix B.

INDEX

Amorphous boron, 62
Anvils, 52, 71
Austenite, 18
Autofrettage, 26
Axial loading shackle, 55

Barium transitions, 4, 67
Belt apparatus, 123, 128
  diamond synthesis, 138
  gaskets and cell assemblies, 131
  high compression, 132
  piston and die, 128
Beryllium copper, 22, 142, 154
Bismuth transitions, 4, 6, 7
Boron, amorphous, 62
Boron, wafer for x-ray diffraction, 105
Boron nitride, 3, 105
Bourdon tube gauge, 38
Boyd and England, piston and cylinder, 125, 127
  cell assemblies, 126
Bridgman anvils, 52
  calibration, 57
  deformation, 61
  electrolytic cell, 60
  miniature diamond, 61
  pressure gradients, 58
  sample cells for electrical measurements, 56
  unsupported area seal, 29
  x-ray diffraction, 61

C scale, Rockwell, 16
Caesium transitions, 4
Calcium transition, 4
Calibration pressure, 4, 5, 7
  at high temperature, 8
  at low temperature, 153
Carbon disulphide, pressure medium, 35, 163

Carbon sleeve heaters, 94, 136
Carbon tetrachloride transition, 4
Cementite, 18
Cerium transitions, 96, 97
Charpy impact test, 17
Chester & Jones low temperature clamp, 146
Coesite-quartz phase diagram, 59
Construction materials, 15
Creep, 18
Cubic anvil apparatus, 88, 107
  calibration, 110
  Japanese, 116
  optical experiments, 112
  Von Platen, 117
Cylinders,
  Lamé criterion, 26
  mean diameter formula, 26
  multi-ring support, 124
  strength, 24
  thick walled, 25
  thin walled, 24
  Von Mises criterion, 26

Dead weight gauge, 38
Delta rings, 30
Diamond synthesis, 1, 99, 138
Diamond windows, 110
  optical transmission, 111
Differential thermal analysis, 136
Differential thermal conductivity analysis, 100, 136
Drickamer supported anvil apparatus, 71 et seq.
  calibration to 500 kb, 75
  for electrical measurements, 71
  Mössbauer resonance, 81
  nuclear magnetic resonance, 137
  optical measurements, 76
  press, 74
  window forming, 77

Drickamer supported anvil apparatus—*cont.*
work hardening, 73
x-ray diffraction, 82
Ductility, 17

Elastic modulus, 17
Electrical connections, 40
Electrical measurements under pressure
belt, 137
cubic anvil, 110
opposed anvils, 56
piston and cylinder, 126
tetrahedral anvil, 94
Electrical resistance pressure gauges, 39
Electron spin resonance, 157

Far infra-red measurements at high pressure, 113
Fatigue, 17
Ferrite, 18
Fourier transform spectroscopy, 113
Free piston gauge, 37
Further reading, 14

Girdle apparatus, 123, 133
Gold chrome pressure manometer, 40
Graphite, heaters, 94, 136

Hardness, steels, 15
High temperature apparatus optical experiments, 79
Honing, 27
Hydraulic intensifier, 36
Hydrostatic pressure
measurement, 37
media, 34

Ice-bomb high pressure device, 144
Indium antimonide, transition, 144
Individual ram multi-anvil apparatus, 101

Interference
in Bridgman anvils, 54
in multi-ring cylinders, 124
Ipatieff seal, 29
Iron-carbon phase diagram, 18
Iron transition, 4
Izod impact test, 17

Jamieson and Lawson x-ray apparatus, 61

Lapping, 27
Lead transition, 4
Line of contact seal, 29
Lithium hydride, 84, 105
Load generation, 35
Low temperature
apparatus, 142
calibration, 153
optical apparatus, 149
pressure media, 143

Magnetic measurements in piston and cylinders, 139
Magnetic resonance, 154
Manganin pressure manometer, 39
Maraging steel, 21
Martensite, 19
Massive support, 52
Mercury, freezing under pressure, 4
Miscellaneous devices, 141
Mössbauer resonance, 81
Multi-anvil apparatus, 87
Multi-ring piston and cylinders, 124

National Bureau of Standards tetrahedral anvil device, 89
National Physical Laboratory x-ray device, 63
Near infra-red optical experiments, 112
Neutron diffraction apparatus, 162
Non-ferrous alloys, 22
Non-magnetic pressure vessels, 154
Nuclear magnetic resonance, 155
Nylon plug seal, 30

INDEX

O-ring seals, 30
Opposed anvil apparatus, 52
Optical energy gaps in II-VI and III-V semiconductors, 81
Optical high pressure devices, 45, 76, 79, 80, 107, 150
Optical transmission of high pressure windows, 111

Patterson seal, 30
Pearlite, 19
Permanent closures, 32
Phase diagram, silicon and germanium, 8
Phase transitions
  by x-ray diffraction, 68
  in II-VI and III-V compounds, 68
  for calibration, 4, 7
Philips high pressure apparatus, 118
Piermarini & Weir x-ray device, 68
Pipestone, 56
Piping connections, 33
Pistons, 28
Piston and cylinder
  above 20 kb, 48
  diamond, 61
  low temperature, 143
  multi ring support, 123
  neutron diffraction, 162
  single stage, 28
  two stage, 52, 123
Piston displacement at low temperatures, 150
Poulter seal, 29
Precompression low temperature devices, 144, 145
Presses, 36
Pressure measurement, 3
Pyrophyllite, 3, 56, 170

Quartz-coesite, transition, 59
Quartz, single crystal optical transmission, 111
Quenching, 19

Rockwell hardness C scale, 16
Rockwell indentors, 63
Rubidium transition, 4

Sapphire windows, 46
  optical transmission, 111
Sealing, 28
Semiconducting compounds, optical transitions, 81, 113
Silica glass densification, 59
Silver chloride, 56, 93
Single crystals under pressure, 71
Sodium chloride,
  for calibration, 106
  windows, 76
Steels
  low carbon content, 18, 168
  low temperature behaviour, 142
  maraging, 21, 169
  nickel alloy, 19, 168
  stainless, 21, 142, 169
  tool, 20, 168
Strength, cylinders, 26

Talc, 126
Tempering, 19
Tetrahedral anvil apparatus, 87, 90, 92
  anvil lifetime, 93
  anvil material, 96
  calibration, 93, 96, 99, 106
  cell assemblies, 94, 105
  diamond synthesis, 99
  gasket material, 96, 105
  heating, 97
  hysteresis, 95
  operation, 92
  pressure distribution, 93
  x-ray diffraction, 103
Thallium transition, 4, 6, 7
Thermocouples
  effect of pressure, 9
  sheathed, 98
Tin transition, 4
Tungsten carbide, 22, 169

175

Units, 1

Valves, 34

Wedge-reaction loading, 88, 91
Weir & Van Valkenberg optical
    apparatus, 69

Windows, optical, 43, 111

X-ray diffraction apparatus, 61, 64,
    103, 68, 63

Yoke for Bridgman anvils, 55
Young's modulus, 17, 168, 169